Natural Computing Series

For further volumes:
http://www.springer.com/series/4190

Gisele L. Pappa • Alex A. Freitas

Automating the Design of Data Mining Algorithms

An Evolutionary Computation Approach

Dr. Gisele L. Pappa
Dept. de Ciência da Computação
Universidade Federal de Minas Gerais
Belo Horizonte
Brazil
glpappa@dcc.ufmg.br

Dr. Alex A. Freitas
School of Computing
University of Kent
Canterbury
UK
a.a.freitas@kent.ac.uk

ISSN 1619-7127
ISBN 978-3-642-26125-1
e-ISBN 978-3-642-02541-9
DOI 10.1007/978-3-642-02541-9
Springer Heidelberg Dordrecht London New York

ACM Computing Classification (1998): I.2.6

Softcover reprint of the hardcover 1st edition 2010

Cover design: KünkelLopka GmbH, Heidelberg

Printed on acid-free paper

Springer is part of Springer Science+Business Media (www.springer.com)

This book is dedicated to all the people who believe that learning is not only one of the most necessary but also one of the noblest human activities.

Preface

Data mining is a very active research area with many successful real-world applications. It consists of a set of concepts and methods used to extract interesting or useful knowledge (or patterns) from real-world datasets, providing valuable support for decision making in industry, business, government, and science.

Although there are already many types of data mining algorithms available in the literature, it is still difficult for users to choose the best possible data mining algorithm for their particular data mining problem. In addition, data mining algorithms have been manually designed; therefore they incorporate human biases and preferences.

This book proposes a new approach to the design of data mining algorithms. Instead of relying on the slow and ad hoc process of manual algorithm design, this book proposes systematically automating the design of data mining algorithms with an evolutionary computation approach. More precisely, we propose a genetic programming system (a type of evolutionary computation method that evolves computer programs) to automate the design of rule induction algorithms, a type of classification method that discovers a set of classification rules from data. We focus on genetic programming in this book because it is the paradigmatic type of machine learning method for automating the generation of programs and because it has the advantage of performing a global search in the space of candidate solutions (data mining algorithms in our case), but in principle other types of search methods for this task could be investigated in the future.

This new approach opens up some exciting avenues for the research and practice of data mining. First of all, once the process of designing a data mining algorithm – normally considered a process requiring a lot of human creativity – has been automated, researchers and practitioners can benefit from a much faster creation of new data mining algorithms. More importantly, the proposed genetic programming system can be used to create rule induction algorithms tailored to the target application domain or the dataset being mined. That is, users are no longer limited to trying to select the best existing algorithm tailored to their data; they can simply ask the computer to automatically generate a new data mining algorithm tailored to their data. It is also interesting to compare automatically designed data mining algorithms with human-designed ones, since findings derived from this comparison can potentially advance the research related to data mining algorithm design.

This is a research-oriented book, and so it is particularly recommended for researchers and postgraduate students in the areas of data mining and evolutionary computation; but we hope it will also provide some useful ideas for data mining practitioners in general. We also hope this book will stimulate further research in the areas of data mining and evolutionary computation.

The book is organized as follows. First, the Introduction explains the motivation for automating the design of data mining algorithms and presents an overview of the system proposed for this task. Next, the book contains two chapters with an overview of data mining and evolutionary computation methods (Chapters 2 and 3 respectively), as well as a chapter with a discussion of research projects related to the topics of automated algorithm design in data mining and optimization (Chapter 4). These three chapters focus on concepts and methods of data mining, evolutionary computation, and optimization that are particularly useful for a better understanding of the new system proposed in the book. The main contribution of the book, a new genetic programming system to automate the design of rule induction algorithms, is described in detail in Chapter 5, and Chapter 6 reports the results of computational experiments evaluating the effectiveness of the proposed system. Finally, Chapter 7 discusses future directions for this emerging area of automation of the design of data mining algorithms.

The research that led to the writing of this book was carried out mainly at the University of Kent, UK, but the last part of the research (including the writing of the book) was carried out both at Kent and at the Federal University of Minas Gerais (UFMG), Brazil. The authors thank their colleagues in both institutions for creating a productive, friendly research environment and for useful discussions about data mining with evolutionary computation methods.

The first author gratefully acknowledges financial support for this research provided by two Brazilian research support agencies, namely CAPES (process number 165002-5) and FAPEMIG.

Belo Horizonte, Brazil, and Canterbury, UK, February 2009

Gisele L. Pappa
Alex A. Freitas

Contents

Acronyms

BNF	Backus–Naur Form
CFG	Context-Free Grammars
EA	Evolutionary Algorithm
EDA	Estimation of Distribution Algorithm
GA	Genetic Algorithm
GGP	Grammar-based Genetic Programming
GGP-RI	Grammar-based Genetic Programming-evolved Rule Induction algorithm
GHC	Grammar-based Hill Climbing
GP	Genetic Programming
GPCR	G-Protein-Coupled Receptor
HC	Hill Climbing
ILP	Inductive Logic Programming
MOGGP	Multiobjective Grammar-based Genetic Programming
MOGGP-RI	Multiobjective Grammar-based Genetic Programming-evolved Rule Induction algorithm
NT	Non-terminal
RI	Rule Induction algorithm
STGP	Strongly Typed Genetic Programming
TAG	Tree-Adjoining Grammar

Chapter 1
Introduction

Data mining is a research area where computational methods are designed and used to extract useful or interesting knowledge (or patterns) from real-world datasets. It is motivated by the fact that the technology for generating and storing data has been improving very fast in the last couple of decades, leading to larger and larger amounts of data stored in computer systems, but our understanding of the corresponding data and the discovery of the knowledge or patterns hidden in such a large amount of data has being growing at a much slower rate.

Data mining can be used to solve a variety of tasks, including classification, regression, clustering, and association discovery. This book focuses on the classification task, which has been extensively studied by researchers and practitioners. In essence, in this task a classification algorithm is given as input a set of preclassified examples (records, data instances). Each example is described by a set of predictor attributes (features) and a special attribute called the class attribute. The classification algorithm has to discover a certain type of predictive relationship (a classification model) between predictor attributes and classes. More precisely, the classification model has to be able to predict the class of an example given the values of the predictor attributes for that example.

Regardless of the data mining task being tackled, one of the main challenges for data mining researchers and users is to choose which algorithm is the best to solve the problem at hand. It is well known that no algorithm is the best across all application domains, and a whole research area named meta-learning [4] was created to investigate this problem. Given this algorithm selection problem, this book proposes to solve it in a rather different way: by automatically constructing data mining classification algorithms, instead of selecting them. In order to achieve that, we turn to one of the most appealing areas of artificial intelligence: the semiautomatic design of computer programs.

The human desire to automatically create computer programs for machine learning tasks dates back to the pioneering work of Samuel in the 1960s, when the term machine learning, meaning "computers programming themselves", was first coined. As years went by, machine learning gained a new definition, related to the system's capability of learning from experience [17]. However, the idea of automatically gen-

G.L. Pappa, A.A. Freitas, *Automating the Design of Data Mining Algorithms*, Natural Computing Series, DOI 10.1007/978-3-642-02541-9_1,

erating computer programs persisted. In the early 1990s, a whole new area dedicated to the study of this idea started to be widely disseminated: genetic programming [12].

The main idea behind genetic programming (GP) is to automatically evolve computer programs capable of producing good solutions (hopefully better than manually produced solutions) for the target problem.

It is important to point out that, in this book, the terms data mining and machine learning are used interchangeably. Although some researchers try to differentiate them, in this book we take the perspective that these two terms converged to refer to the same broad type of field, involving the analysis of real-world datasets.

This book deals with the design of a specific type of classification algorithms: rule induction algorithms, which are introduced in Section 1.1. Rule induction algorithms will be automatically designed using an evolutionary algorithm, more precisely, a genetic programming algorithm, which is presented in Section 1.2. The motivations to automatically design a rule induction algorithm, including dealing with the problem of selecting a suitable classification algorithm to a dataset, avoiding human biases, and introducing a whole new level of automation to data mining tasks, are discussed in Section 1.3. Finally, Section 1.4 gives an overview of the proposed system.

1.1 Rule Induction Algorithms

The basic idea of using evolutionary computation to automate the design of a data mining algorithm is very generic, and in practice, to make the task more tractable, we have to focus on a specific type of data mining algorithm. In this work we focus on rule induction algorithms, which will be discussed in detail in Chapter 2. The main reason to focus on rule induction algorithms is that this type of algorithm has the advantage of discovering knowledge expressed in the form of *if-then* rules that are intuitively comprehensible to users. More precisely, rule induction algorithms discover rules in the form *if* (conditions) *then* (predicted class), where the rule's consequent (*then* part) predicts a class for any example (record, data instance) that satisfies the conditions included in the rule's antecedent (*if* part). For instance, the rule *if* (*Salary* = *high*) and (*C/A-balance* = *high*) *then* (*Credit* = *good*) predicts that any costumer with a *high salary* and a *high current account balance* belongs to the class of *good credit* customers.

The fact that the discovered knowledge is expressed in a form that is easily understandable by users is an advantage of rule induction algorithms that is important in many applications. In particular, knowledge comprehensibility tends to be important in applications where the discovered knowledge should be interpreted by the user, either to support the user's important decisions about the application domain (rather than letting the machine make an important decision) or to give the user more insight about the data and the application domain.

For instance, in the application domain of medicine, rules predicting whether or not a patient has a certain disease should be interpreted by a medical doctor, who could use them to support (together with other sources of evidence) a decision about a treatment to be prescribed to the patient. Note that in general it would not be acceptable to let a computer make an automated decision about a patient's treatment without involving a medical doctor, since human lives are at stake and computers have no understanding of the meaning of discovered knowledge. Even if the discovered rules are not used to support decisions related to the treatment of individual patients, the analysis of discovered rules can give medical doctors new insights about the application domain – as reported for instance in [29], where the system found classification rules about scoliosis that were found interesting and gave new insights to medical doctors.

As another example, in the financial application domain, Dhar et al. [6] remark that in their experience decision makers are more likely to invest capital in cases where the knowledge or patterns discovered by the data mining algorithm are easy to understand. They point out, in particular, that "Rules generated by pattern discovery algorithms are particularly appealing... because they can make explicit to the decision maker the particular interactions among the various market indicators that produce desirable results."

Knowledge comprehensibility is also important in scientific discovery, i.e., where the data being mined comes from a given field of science and the discovered knowledge will be interpreted by scientists, who hope to get more insight about the data and the underlying field of science. In particular, the importance of the comprehensibility of discovered knowledge in bioinformatics – more precisely, in the prediction of protein functions with data mining methods – is discussed in detail in [9].

It should be noted that rule induction algorithms are not the only kind of classification algorithms capable of discovering comprehensible knowledge. For instance, decision tree induction algorithms also can procedure comprehensible knowledge, in the form of a decision tree [3, 26], which has the advantage of being a graphical representation. On the other hand, *if-then* classification rules have the advantage of being a more modular knowledge representation, since each rule can often be interpreted by the user independently from other rules, and it is often the case that a decision tree can be simplified into a relatively smaller, simpler set of classification rules [25, 26].

In any case, as mentioned earlier, in this book we propose an evolutionary computation method for automating the design of rule induction algorithms. Proposing another evolutionary method for automating the design of decision tree induction algorithms (or another kind of data mining algorithm, for that matter) would be an interesting direction for future research.

1.2 Evolutionary Computation

Evolutionary computation is an area of computer science which considers using abstractions of concepts and principles of natural evolution – in particular, the Darwinian principle of natural selection – as an inspiration or metaphor for the design of computational methods. Algorithms following the principle of natural selection are broadly referred to as Evolutionary Algorithms (EAs).

Since EAs are very generic and very flexible computational methods, they can be used in a number of different applications, for different purposes. In this book we are interested in using an EA for finding the best possible solution for a specific computational problem, namely the problem of automatically designing and implementing new rule induction algorithms. It should be noted that, unlike other applications of EAs, such as simulating or modeling a complex biological process or creating artificial life, our specific target problem is associated with some well-defined measures of performance. In particular, we want to automatically design rule induction algorithms that have a predictive accuracy as high as possible when applied to real-world classification datasets. (Other desirable criteria to be satisfied by the automatically designed rule induction algorithms will be discussed later, but their predictive accuracy is usually considered the most important criterion, and is enough for the purposes of the discussion in this section.)

In this context, the EA should be designed from an "engineering perspective," i.e., the EA should be designed to be an effective method for solving the target problem. This suggests that, in addition to using abstractions and inspiration from natural selection, the design of the EA should also incorporate some knowledge about the type of target problem to be solved. In this case, the extent to which the EA is mimicking real biological principles is relatively less important. To examine a well-known example about the importance of both biological inspiration and problem-specific knowledge in engineering, consider the design of airplanes. Observing the flights of birds in nature has been clearly useful in the design of airplanes, since human-designed airplanes have artificial wings inspired by birds' wings. However, unlike birds, human-designed airplanes do not fly by flapping their wings; rather, they fly due to the use of mathematical and physical aerodynamics concepts and principles in their design and construction.

1.2.1 Genetic Programming

The specific type of EA used in this book is Genetic Programming (GP), which essentially aims at evolving computer programs. EAs are a broad type of computational problem-solving methods, with many types of evolutionary algorithms available in the literature, and of course there are many other (non-evolutionary) types of computational problem-solving methods available. This raises the question of the justification for using GP – rather than another evolutionary or non-evolutionary method – to automate the design of new rule induction algorithms.

First of all, GP can be considered the paradigmatic method for automatic program evolution, since it has been invented specifically for the purpose of automatic program creation. The focus on creating computer programs is a major distinction between GP and other types of evolutionary algorithms. In essence, the two main ideas behind this distinction are as follows, and a more detailed discussion of these points will be presented in Chapter 3.

First, GP manipulates candidate solutions (for a given target problem) that have a considerably greater expressiveness power than the candidate solutions manipulated by other types of evolutionary algorithms. In particular, GP can manipulate candidate solutions consisting of not only data and variable values, but also functions and operators and potentially complex programming elements such as loop statements. In contrast, in general other types of evolutionary algorithms manipulate solutions consisting of only data and variable values, and not including any kind of executable structure.

Secondly, GP algorithms aim (or at least should aim) at finding a generic solution for the target problem, i.e., a solution in the form of a generic program that could be used to solve any particular instance of that problem. In contrast, other types of evolutionary algorithm typically focus on finding a solution for a specific instance of the target problem. For example, consider the very well-known Traveling Salesman Problem, where a salesman starting in a given city has to find the shortest tour passing exactly once through each other city and then returning to the starting city. A GP algorithm for this problem finds a generic program capable of solving this problem for any given set of cities and distances between cities. To find such a generic program, the GP is given as input a number of different instances of the problem, each instance with a different set of cities and distances between cities. In contrast, other types of evolutionary algorithm aim at finding the optimal tour for just one particular instance of the problem, consisting of just one particular set of cities and distances between cities given as input to the evolutionary algorithm.

An additional motivation to focus on GP in this work is that, like other evolutionary algorithms, GP is a global search method, which iteratively evolves a population of candidate solutions (programs) spread across different regions of the search space. Hence, GP is in general less likely to get trapped in local optima in the search space than local search methods, which typically work with just one candidate solution at a time.

Furthermore, GP has been extensively used by researchers and practitioners. A very large body of literature on GP is available to guide us in the design of a GP algorithm, such as several books [1, 12, 13, 24] and a large number of research papers in proceedings of conferences on evolutionary algorithms. Also, Koza et al. [13] mention a number of problems where GP found a solution that was considered competitive with the best solution produced by human beings, according to the authors' specific human-competitiveness criteria.

Despite the above motivations, there are some caveats to be considered in the use of GP for automating the design of a data mining algorithm, as follows. First, to the best of our knowledge, before the research reported in this book was carried out, GP had never been used to automate the design of a full rule induction algorithm, and there are few previous works on automating the design of data mining algorithms in general – as will be discussed in Chapter 4.

In addition, although the literature on GP is very large as mentioned earlier, in practice unfortunately the vast majority of that literature is using GP to evolve solutions that are "programs" only in a loose sense of the word. The evolved solutions usually have some "executable structures," like mathematical functions, but they usually do not have more complex and fundamental program constructs such as loop statements. This is because in the vast majority of the literature GP has been applied to relatively simple benchmarking problems, where a full program with loops and more complex program constructs is not needed. The task of creating a new data mining algorithm can be considered much more ambitious and challenging than the benchmarking tasks typically addressed in the GP literature. This point will be discussed in detail in Chapter 3. Hence, a straightforward application of a conventional GP algorithm would be unlikely to lead to the automatic design of full rule induction algorithms.

Moreover, recall that when using an evolutionary algorithm for solving a real-world problem from an engineering perspective it is important to incorporate into the evolutionary algorithm specific knowledge about the type of problem being solved. In our case this is done by using a grammar, so that the proposed GP system is actually a grammar-based GP system. The grammar incorporates background knowledge about the structure of rule induction algorithms and is instrumental in mitigating difficult problems associated with the automated creation of programs, such as how to avoid infinite loops.

To summarize, although automatic program creation can be achieved by using in principle any other generic computational search method (with suitable adaptations to allow them to output programs), and despite the fact that there is very little related work on GP for automating the design of a data mining algorithm, the fundamental and paradigmatic nature of GP as an automated program creation method makes GP a natural choice for the type of search method used in the research reported in this book.

At this point it is important to mention that there has also been a large amount of research in the area of Inductive Logic Programming (ILP) [2, 7, 15]. The basic idea of ILP methods is to learn (or induce), from examples of and background knowledge about the underlying application domain, a kind of logic program that is a model of a certain type of relationships between attributes (or variables) in that application domain. Note that so far ILP has been used to produce classification models (or models for other types of data mining tasks), rather than to produce a full classification algorithm – a detailed discussion between classification models and classification algorithms will be presented in Section 4.2. Also, ILP algorithms tend to use a particular type of programming representation, based on the Prolog programming language, while GP is arguably associated with a more flexible program representation, which can be adjusted by the user in a relatively easy way by defining appropriate symbols in the function and terminal sets, as will be discussed in Chapter 3. Nonetheless, creating an ILP system for automatically designing a data mining algorithm would also be a valid and interesting research direction.

It should be noted that we do not claim that GP is "the best" type of method for automating the design of data mining algorithms. The problem of finding the best type of method for this task is very open, with GP being a natural baseline solution. Much more research will be needed to determine if GP or another type of method is more suitable for this task, since this is a research topic that is in its infancy.

In any case, it is worth mentioning that in this book we report computational results that provide a preliminary evaluation of the effectiveness of GP for the task of automating the design of a rule induction algorithm in comparison with a simpler type of search method, namely a local search hill-climbing method that has been suitably adapted for the same task. In particular, both the proposed GP method and the hill-climbing method used the same grammar to guide their search for new rule induction algorithms, to make the comparison between them fair. Computational results reported in Chapter 6 show that the GP method has, overall, created better rule induction algorithms (i.e., algorithms with a higher predictive accuracy in general) than the hill-climbing method.

1.3 The Motivation for Automating the Design of Classification Algorithms

The motivation for automating the design of classification algorithms is threefold. The first two of these motivations, i.e., addressing the problem of selective superiority of classification algorithms and the human bias on the design of these algorithms, are mostly technical. Our last motivation, however, is more philosophical, and discusses a new level of automation in data mining from the perspective of artificial intelligence systems. These three points are discussed in the next three subsections.

1.3.1 The Problem of the Selective Superiority of Classification Algorithms

It is well known that there is no classification algorithm which is the "best" across all application domains or datasets. This has been shown both theoretically and empirically [4, 16, 18, 27, 28]. This raises the question of how to find the best classification algorithm for a given target dataset, a form of the "selective superiority problem" [5] for classification algorithms. The solution proposed in this book is to automatically create a classification algorithm tailored to the target dataset, but before elaborating on this, let us briefly review other solutions proposed in the literature for this problem.

One approach for finding the best classification algorithm for a given target dataset consists of selecting the best algorithm out of a list of existing algorithms. A straightforward way of implementing this approach consists of trying to run several different candidate classification algorithms on the target dataset, measuring the predictive accuracy obtained by each of them, and simply choosing the algorithm that obtained the best accuracy. Clearly, this approach is not elegant, being an ad-hoc approach based on trial and error. In addition, it has the disadvantage of being computationally slow (requiring runs of many candidate classification algorithms), a problem aggravated by the fact that it ignores lessons learned from previous runs of the algorithms: every time that we have a new dataset, we need to run all the candidate classification algorithms again.

An improvement – at least from a scientific point of view – over the above straightforward approach consists of using a meta-learning approach. Meta-learning is a large subarea of machine learning research by itself [4], and it will be briefly reviewed in Section 2.5. Here, we briefly discuss the basic idea of meta-learning as an approach for algorithm selection. In this approach many different (base-level) classification algorithms are run on many different (base-level) datasets and the results – in particular, predictive accuracies or related measures – of this "base-level learning" are recorded. These results are then analyzed by a "meta-learning" algorithm. This often takes the form of a meta-classification algorithm, which creates a meta-classification model, i.e., a model that predicts which of the base-level classification algorithms (which "meta-class") will be the best algorithm for a given base-level dataset, based on "meta-attributes" describing the characteristics of that dataset. Note that meta-attributes describe an entire dataset, unlike base-level attributes, which describe just an example (record) in a dataset. For instance, three very simple meta-attributes could be the number of examples, the number of attributes, and the number of classes in the base-level dataset, but much more complex meta-attributes can be used.

Despite significant progress in meta-learning research, meta-learning remains a difficult problem, and it is still used relatively little in practice. This seems due, at least in part, to two limitations of meta-learning. First, classification is an extremely generic data mining task. Classification algorithms can be applied to any application domain where a suitable dataset is available, involving an extremely large number of different types of datasets. In particular, classification datasets vary enormously in the number of examples, the number of attributes, the degree of correlation between the attributes and the classes, the degree of attribute interaction, and a number of other dataset characteristics that interact in a highly nonlinear way. Thus, it is difficult to find a good set of meta-attributes that capture the relevant properties of so many different types of datasets for meta-learning purposes.

Secondly, conventional meta-learning methods perform just algorithm selection, i.e., they try to find the best possible classification algorithm for a target dataset from a limited and predefined set of existing classification algorithms. In principle there is no guarantee that this predefined (and usually small) set of candidate classification algorithms includes a good (let alone the best or near best) classification algorithm for the target dataset.

The proposed idea of automatically creating a new data mining algorithm essentially bypasses the two aforementioned drawbacks or limitations of conventional meta-learning. First, the proposed approach avoids the difficult problem of identifying meta-attributes representing relevant characteristics of a number of very diverse types of datasets. Secondly, the proposed approach directly constructs a new classification algorithm, which is a more flexible approach that is not limited to just selecting a classification algorithm from a set of predefined ones.

In particular, the construction of a new classification algorithm can be done in a way tailored to the target dataset. To see the appeal behind this idea, let us make an analogy with the process of buying clothes of the "right size" for a person. Most people choose their clothes' size out of a small list of predefined sizes. This is a simple and relatively cheap solution, since the mass production of clothes in standard sizes helps to reduce the costs of the clothes. The disadvantage of this approach is that it is not easy for many people to find a standard size very suitable to them. For some people one standard size of some trousers might have the right size in the waist but be too long or too short in the legs, or have related problems. In principle a more effective solution for a person is to hire a tailor to make clothes exactly according to his or her body's measures. In this case the clothes will in principle fit better to the person's body. In the clothes industry, this approach is less popular mainly because it is considerably more expensive, since tailor-made clothes involve the manual skill of a tailor, in contrast with the industrialized mass production of clothes in a small set of standard sizes.

Now, back to the problem of finding the best classification algorithm for a given dataset. The standard meta-learning approach of selecting the best classification algorithm out of a list of predefined algorithms is like that of buying clothes of some standard size in the above analogy. The approach proposed in this book, to automate the design of classification algorithms, is like that of hiring a tailor to make clothes of a person's right size, as determined by his or her body's measures. Intuitively, the latter tends to be more effective, but it has the disadvantage of being more expensive in conventional data mining. After all, in conventional data mining, it would not be feasible to hire a data mining expert to design a classification algorithm tailored to each dataset, since the number of different datasets and application domains seems vastly greater than the number of experts in classification algorithm design, and the manual process of algorithm design is very time consuming and expensive.

However, if the process of designing a classification algorithm is automated as proposed in this book, the cost and time of such algorithm design is significantly reduced, and in principle it becomes feasible to have a classification algorithm automatically designed for each new dataset provided by a user. An important question, of course, is whether the automatically designed algorithms will be effective in the sense of being better or at least competitive with manually designed algorithms, and computational results reported in Chapter 6 will show some evidence for the effectiveness of the proposed approach.

1.3.2 Human Biases in Manually Designed Algorithms

Designing a classification algorithm is not an easy task, and it is normally done by a researcher who is an expert in classification. Such a manual design of classification algorithms is naturally slow and costly, and subject to human biases and preferences.

Consider, for instance, rule induction algorithms, the type of classification algorithm that is the focus of this book. The vast majority of rule induction algorithms are "greedy" in the sense that they typically build a classification rule incrementally, by adding or removing one condition (attribute-value pair) at a time to or from a partial classification rule. This greediness makes them have trouble with datasets with strong attribute interactions. In essence, attribute interaction occurs when some attributes seem to have low predictive power (i.e., their values have little correlation with the classes to be predicted) by themselves, but they actually have high predictive power when they are considered together with other attributes. See [8] for a review of the importance of attribute interaction in data mining and [6] for a discussion of attribute interaction particularly in the context of evolutionary algorithms for discovering classification rules.

The greediness of the vast majority of rule induction algorithms can be regarded, at least in part, as a consequence of the human biases imposed on such algorithms by their designers. Actually, the greedy search performed by some rule induction algorithms seems conceptually similar, at a high level of abstraction, to the approach used by humans when they learn to form concepts.

How humans learn to form concepts is summarized, for instance, by Gardner [10], who reviewed a classical psychological study on this topic. In that study, subjects were asked to recognize instances of geometric concepts like *large red triangle* or *tall blue cylinder*. A concept (category) was arbitrarily defined and each object shown in a card was considered either a member or a non-member of that concept. Each subject was exposed to one card at a time, asked in each case whether that card was a member of the concept, and then told whether his or her response was correct. The subject's task was to identify the attributes defining the target concept, in order to select all, and only, those cards exhibiting the defining features of the concept. The researchers found that the most foolproof method used by the subjects was *conservative focusing*, where the subject found a positive instance of the concept and then made a sequence of choices, each altering *only a single attribute value* of that first "focus" card and testing whether the change produced a positive or negative instance of the concept. As pointed out by Gardner, that study was quite influential, and it seems that at that time no one realized that the use of such artificial concepts – where categories were unambiguously defined by a small number of defining attribute values – might invalidate some of the findings.

Once we have a method for automatically designing a rule induction algorithm, in principle the machine-designed algorithms could have very different biases from the human-designed algorithms. If such an automated design method is successful, the machine-designed algorithms could have better predictive performance on some kinds of datasets where current human-designed algorithms are not very successful.

Some evidence for the fact that genetic programming can automatically produce rule induction algorithms that show at least some aspects of innovation (though they are not entirely original), by comparison with existing human-designed algorithms, will be shown in Chapter 6. In addition, it seems worth pointing out that there is stronger evidence for the "creative capability" of evolutionary computation in other applications. For instance, Keane and Brown [11] designed a simple satellite dish holder boom – which connects the satellite's body with the dish needed for communication – using an evolutionary algorithm. They gave to the evolutionary algorithm a structure with regularly repeated patterns, and the genetic algorithm modified its geometry by changing the three-dimensional coordinates of each pattern. The resulting designed satellite dish holder boom looks strange to a human engineer, due to the lack of symmetry. Yet, it is approximately 20,000% better than the conventional human-designed shape.

Another example of the creative capability of evolutionary algorithms is the scientific discovery of a new form of boron [20]. Boron is one of the most mysterious chemical elements, and for over a century scientists tried to isolate it. However, given its sensitivity to impurities, it was not until 1909 that a 99% sample of pure boron was produced. In [20], Oganov et al. used an evolutionary algorithm to determine the structure of a newly discovered form of boron. In the same way carbon can be found in the form of graphite or diamond, boron also has different forms. However, to the current date, only four of them are known, where the last was discovered by an evolutionary algorithm.

This last discovery started when scientists found out that, under certain conditions of pressure, the boron was stable. However, they could not determine the molecular structure under these conditions. Hence, this job was left to the evolutionary algorithm. Given a number of crystal structures, the evolutionary algorithm was used to calculate the energy necessary to hold the structure together. Then, the structures went through crossover and mutation operations, and evolution was carried out until the crystal converged to a stable form of boron. The newly discovered stable structure is very resistant to heat, and might be attractive for industrial use.

1.3.3 A New Level of Automation in Data Mining

Another motivation for automating the design of classification algorithms is somewhat related to artificial intelligence and philosophy of science issues. From an artificial intelligence point of view, creating a method for automating the design of classification algorithms could perhaps be considered a significant step towards more generic and more autonomous artificial intelligence systems. After all, if a computer is capable of automatically creating a new algorithm to solve a target problem, the computer will be less dependent on human programmers. As a caveat, it is fair to mention that the method proposed in this book for automating algorithm design offers a limited contribution towards this goal, since achieving it still depends, strongly, on background knowledge provided by a human expert, in the form of a grammar specifying the structure of rule induction algorithms. In any case, the method seems to be a significant step in the direction of increasing a little the autonomy of a computational system from an artificial intelligence point of view.

From the point of view of issues related to the philosophy of science, one can consider data mining as a kind of science. More precisely, as mentioned earlier, data mining is very closely related to machine learning, and the two terms are used interchangeably in this book. Machine learning has been broadly defined by Mitchell [19] as the study of computer programs that improve their performance at some task through experience. Clearly, the "experience" used for learning is usually represented in the form of a dataset, which naturally leads to the high similarity between data mining's and machine learning's goals and methods. Langley presented a view of machine learning as a science that can be applied to data mining too, from this book's perspective. According to Langley [14, p. 5], "If machine learning is a science, then it is clearly a science of algorithms." With this perspective, the science of machine learning or data mining has the goal of discovering the best possible algorithm for analyzing real-world datasets.

It is interesting to think about exactly what is being automated in conventional data mining. From the point of view of a data mining user, the analysis of his or her data is a process that is at least partially automated by running a data mining algorithm on that data. (We say "partially automated" because in practice the user should still contribute to the data analysis process with his or her background knowledge about the data and the application domain.) However, the actual process of designing a data mining algorithm, i.e., the actual way in which the science of data mining proceeds – by searching for the best possible data mining algorithm – is hardly automated at present. Data mining algorithms are normally manually designed by experts in the corresponding type of algorithm, in a slow and costly process, subject to human biases and preferences, as mentioned earlier.

To summarize the above point, the data mining research community as a whole is working hard to automate the data analysis task of users, but it is not practicing what it preaches, in the sense that it is not automating the design of data mining algorithms. The idea of automating (to some extent) the design of such algorithms, proposed here, corresponds to a new level of automation in data mining, which, if followed up on and improved by many researchers, could lead to a new generation of data mining algorithms.

1.4 Overview of the Proposed Genetic Programming System

When manually designing a rule induction algorithm (or any data mining algorithm), the designer has to make decisions about the main components his or her algorithm will have, implement it, and make it general enough so that it can be later applied to a variety of datasets coming from different application domains. Automatically designing a data mining algorithm from scratch would mean teaching a machine to follow these same steps. However, implementing an algorithm to do

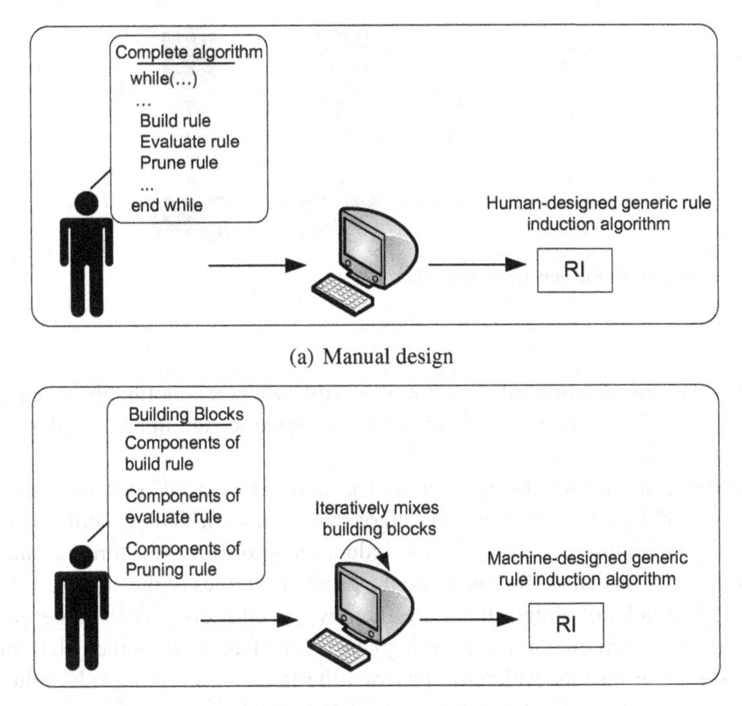

(a) Manual design

(b) Partially-automated design

Fig. 1.1 The design of rule induction algorithms

that from scratch would not be feasible nowadays. Hence, in the system proposed here, a designer creates a set of production rules that are followed when creating a rule induction algorithm [21, 22, 23]. These production rules are then introduced to the system in the form of a grammar, which is used to guide the algorithm design process.

Figure 1.1 shows the differences between the manual and automated rule induction algorithm design approaches. While in Fig. 1.1(a) the human designer creates the complete rule induction algorithm, in Fig. 1.1(b) he or she only identifies its main components, or "building blocks," which are the elements used to define how classification rules can be created, evaluated, or pruned. The building blocks are included in the grammar in order to incorporate background knowledge into the algorithm design process. In a second stage, a computational system is responsible for designing the best possible rule induction algorithm by iteratively combining the building blocks included in the grammar. Note that both the approaches shown in Figs. 1.1(a) and 1.1(b) create a complete rule induction algorithm, which can then be used to discover classification rules for different datasets in different application domains. This is illustrated in Fig. 1.2, where the different shades of gray associated with different datasets denote that they can be very different datasets (leading to

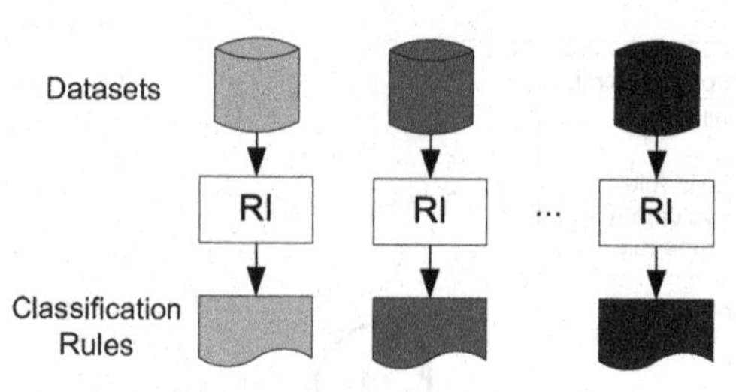

Fig. 1.2 Application of the generated algorithm

very different classification rules, of course), while the box denoting the rule induction (RI) algorithm is the same (denoting a very generic rule induction algorithm) across that figure.

Once the grammar has been defined, a grammar-based GP system is used to automate the design of rule induction algorithms. First, the GP system randomly generates a population of candidate rule induction algorithms (each represented in the form of a tree structure) by following the production rules of the grammar. These candidate rule induction algorithms are then evaluated according to some quality criteria (mainly their predictive accuracy) in a set of datasets named the "meta-training set". During this evaluation process, the trees representing rule induction algorithms are converted into Java code and then compiled into machine code. Then each of those candidate rule induction algorithms is executed on each dataset in the meta-training set.

After evaluation, the values of predictive accuracy of the candidate rule induction algorithms are used to select the best algorithms to undergo operations such as crossover or mutation (loosely analogous to their natural genetics counterpart). These operations produce new candidate rule induction algorithms ("offspring" algorithms) out of the selected "parent" algorithms. This process is repeated until a stopping criterion is met, for example, a maximum number of generations (iterations) is reached. Hence, due to the iterative selection of good-quality rule induction algorithms and their modification by genetic operators, the population of candidate rule induction algorithms is expected to gradually evolve to a near-optimal rule induction algorithm, according to the algorithm quality criterion used to guide the evolutionary search. After the evolutionary process is over, the best rule induction algorithm produced by the GP system is evaluated on a new set of datasets called the "meta-test set." We emphasize that the meta-test set contains only datasets that were not used in the meta-training set. A detailed description of the proposed GP system can be found in Chapter 5.

References

1. Banzhaf, W., Nordin, P., Keller, R.E., Francone, F.D.: Genetic Programming – An Introduction; On the Automatic Evolution of Computer Programs and its Applications. Morgan Kaufmann (1998)
2. Bergadano, F., Gunetti, D.: Inductive Logic Programming: from machine learning to software engineering. MIT Press (1996)
3. Bramer, M.: Principles of Data Mining. Springer (2007)
4. Brazdil, P., Giraud-Carrier, C., Soares, C., Vilalta, R.: Metalearning: applications to data mining. Springer (2009)
5. Broadley, C.: Addressing the selective superiority problem: automatic algorithm/model class selection. In: Proc. of the 10th Int. Conf. on Machine Learning, pp. 17–24. Morgan Kaufmann (1993)
6. Dhar, V., Chou, D., Provost, F.J.: Discovering interesting patterns for investment decision making with GLOWER – a genetic learner overlaid with entropy reduction. Data Mining and Knowledge Discovery **4**(4), 251–280 (2000)
7. Dzeroski, S., Lavrac, N. (eds.): Relational Data Mining. Springer (2001)
8. Freitas, A.A.: Understanding the crucial role of attribute interaction in data mining. Artificial Intelligence Review **16**(3), 177–199 (2001)
9. Freitas, A.A., Wieser, D., Apweiler, R.: On the importance of comprehensible classification models for protein function prediction. IEEE/ACM Transactions on Computational Biology and Bioinformatics (in press)
10. Gardner, H.: The mind's new science: a history of the cognitive revolution. Basic Books, New York (1985)
11. Keane, A., Brown, S.: The design of a satellite boom with enhanced vibration performance using genetic algorithm techniques. In: Proc. of Conf. on Adaptative Computing in Engineering Design and Control, pp. 107–113. Plymouth, UK (1996)
12. Koza, J.R.: Genetic Programming: on the programming of computers by the means of natural selection. The MIT Press, Massachusetts (1992)
13. Koza, J.R., Keane, M.A., Streeter, M.J., Mydlowec, W., Yu, J., Lanza, G.: Genetic Programming IV: Routine Human-Competitive Machine Intelligence. Kluwer Academic Publishers (2003)
14. Langley, P.: Elements of Machine Learning. Morgan Kaufmann (1996)
15. Lavrac, N., Dzeroski, S.: Inductive Logic Programming: techniques and applications. Ellis Horwood (1994)
16. Lim, T., Loh, W., Shih, Y.: A comparison of prediction accuracy, complexity, and training time of thirty-three old and new classification algorithms. Machine Learning **40**(3), 203–228 (2000)
17. Michalski, R.S., Carbonell, T.J., Mitchell, T.M. (eds.): Machine Learning: An Artificial Intelligence Approach. TIOGA Publishing Co., Palo Alto, USA (1983)
18. Michie, D., Spiegelhalter, D.J., Taylor, C.C., Campbell, J. (eds.): Machine learning, neural and statistical classification. Ellis Horwood, Upper Saddle River, NJ, USA (1994)
19. Mitchell, T.: Machine Learning. Mc Graw Hill (1997)
20. Oganov, A.R., Chen, J., Gatti, C., Ma, Y., Ma, Y., Glass, C.W., Liu, Z., Yu, T., Kurakevych, O., Solozhenko, V.L.: Ionic high-pressure form of elemental boron. Nature **457**, 863–867 (2009)
21. Pappa, G.L., Freitas, A.A.: Automatically evolving rule induction algorithms. In: J. Fürnkranz, T. Scheffer, M. Spiliopoulou (eds.) Proc. of the 17th European Conf. on Machine Learning (ECML-06), *Lecture Notes in Computer Science*, vol. 4212, pp. 341–352. Springer Berlin/Heidelberg (2006)
22. Pappa, G.L., Freitas, A.A.: Automatically evolving rule induction algorithms tailored to the prediction of postsynaptic activity in proteins. Intelligent Data Analysis **13**(2), 243–259 (2009)
23. Pappa, G.L., Freitas, A.A.: Evolving rule induction algorithms with multi-objective grammar-based genetic programming. Knowledge and Information Systems **19**(3), 283–309 (2009)

24. Poli, R., Langdon, W., McPhee, N.: A Field Guide to Genetic Programming. Freely available at http://www.gp-guide.org.uk (2008)
25. Quinlan, J.R.: Simplifying decision trees. International Journal of Man-Machine Studies **27**, 221–234 (1987)
26. Quinlan, J.R.: C4.5: programs for machine learning. Morgan Kaufmann (1993)
27. Rao, R.B., Gordon, D., Spears, W.: For every generalization action, is there really an equal and opposite reaction? Analysis of the conservation law for generalization performance. In: Proc. of the 12th Int. Conf. on Machine Learning (ICML-95), pp. 471–479. Morgan Kaufmann (1995)
28. Schaffer, C.: A conservation law for generalization performance. In: Proc. of the 11th Int. Conf. on Machine Learning (ICML-94), pp. 259–265. Morgan Kaufmann (1994)
29. Wong, M.L., Leung, K.S.: Data Mining Using Grammar-Based Genetic Programming and Applications. Kluwer, Norwell, MA, USA (2000)

Chapter 2
Data Mining

2.1 Introduction

Data mining consists of automatically extracting hidden knowledge (or patterns) from real-world datasets, where the discovered knowledge is ideally accurate, comprehensible, and interesting to the user. It is an interdisciplinary field and a very broad and active research area, with many annual international conferences and several academic journals entirely devoted to the field. Therefore, it is impossible to review the entire field in this single chapter.

Data mining methods can be used to perform a variety of tasks, including association discovery, clustering, regression, and classification [30, 32]. In this book, we are particularly interested in the classification task of data mining, which is the main subject of this chapter. When performing classification, a data mining algorithm generates a classification model from a dataset, and this model can be later applied to classify objects whose classes are unknown, as explained in Section 2.2.

The classification models generated by data mining algorithms can be represented using a variety of knowledge representations [6, 78]. Broadly speaking, classification models can be divided into two categories, according to the type of knowledge representation being used: human-comprehensible and black-box models. Examples of human-comprehensible models include classification rules, decision trees, and Bayesian networks, whereas artificial neural networks and support vector machines represent black-box models.

The focus of this book is on human-comprehensible models, especially rule-based models. In this context, Section 2.3 introduces decision tree induction, a type of classification algorithm that generates classification models in the form of a decision tree, which can be easily transformed into a set of classification rules. Section 2.4, in turn, presents a detailed description of rule induction algorithms following the sequential covering approach, which are the type of classification algorithms whose automated design is proposed in this book, as will be discussed in Chapter 5.

After describing these two most common approaches used to generate rule-based classification models, Section 2.5 presents a discussion on meta-learning methods.

G.L. Pappa, A.A. Freitas, *Automating the Design of Data Mining Algorithms*,
Natural Computing Series, DOI 10.1007/978-3-642-02541-9_2,

The area of meta-learning appeared as an alternative to help in choosing appropriate classification algorithms for specific datasets, as it is well known that no classification algorithm will perform well in all datasets. Finally, Section 2.6 summarizes the chapter.

2.2 The Classification Task of Data Mining

This section provides an overview of basic concepts and issues involved in the classification task of data mining. A more detailed discussion can be found in several good books about the subject, including [41] and [76].

Before defining the classification task, it is important to emphasize that, broadly speaking, data mining algorithms can follow three different learning approaches: supervised, unsupervised, or semi-supervised. In supervised learning, the algorithm works with a set of examples whose labels are known. The labels can be nominal values in the case of the classification task, or numerical (continuous) values in the case of the regression task. In unsupervised learning, in contrast, the labels of the examples in the dataset are unknown, and the algorithm typically aims at grouping examples according to the similarity of their attribute values, characterizing a clustering task. Finally, semi-supervised learning is usually used when a small subset of labeled (pre-classified) examples is available, together with a large number of unlabeled examples. Here, we are dealing with supervised learning.

Therefore, in the (supervised) classification task each example (record) belongs to a class, which is indicated by the value of a special goal attribute (sometimes called the target attribute, or simply the class attribute). The goal attribute can take on categorical (nominal) values, each of them corresponding to a class. Each example consists of two parts, namely a set of predictor attribute (feature) values and a goal attribute value. The former are used to predict the value of the latter. Note that the predictor attributes should be relevant for predicting the class (goal attribute value) of an example.

At this point a brief note about terminology seems relevant. Since the classification task is studied in many different disciplines, different authors often use different terms to refer to the same basic elements of this task. For instance, an example can be called a data instance, an object, a case, or a record. An attribute can be called a variable or a feature. In this book we will use mainly the terms example and attribute.

In the classification task the set of examples being mined is divided into two mutually exclusive and exhaustive (sub)sets, called the training set and the test set, as shown in Fig. 2.1. The classification process is correspondingly divided into two phases: training, when a classification model is built (or induced) from the training set, and testing, when the model is evaluated on the test set (unseen during training).

In the training phase the algorithm has access to the values of both predictor attributes and the goal attribute for all examples of the training set, and it uses that information to build a classification model. This model represents classifica-

Training set
Known-class examples

	...		class
			good
			bad
			bad
	...		good
			good
			bad

Test set
Unknown-class examples

	...		class
			?
			?
			?
	...		?
			?
			?

Fig. 2.1 Data partitioning for classification

tion knowledge – essentially, a relationship between predictor attribute values and classes – that allows the prediction of the class of an example given its predictor attribute values.

Note that, from the viewpoint of the classification algorithm, the test set contains unknown-class examples. In the testing phase, only after a prediction is made is the algorithm allowed to "see" the actual class of the just-classified example. One of the major goals of a classification algorithm is to maximize the predictive accuracy (generalization ability) obtained by the classification model when classifying examples in the test set unseen during training.

The knowledge (classification model) discovered by a classification algorithm can be expressed in many different ways. As mentioned before, in this book we are mainly interested in the discovery of high-level, easy-to-interpret classification rules of the form *if* (conditions) *then* (class), where the rule antecedent (the *if* part) specifies a set of conditions referring to predictor attribute values, and the rule consequent specifies the class predicted by the rule to any example that satisfies the conditions in the rule antecedent. A simple example of a classification rule would be *if (Salary >100,000 euros* and *Has_debt = no) then (Credit = good)*.

These rules can be generated using different classification algorithms, the most well known being the decision tree induction algorithms and sequential covering rule induction algorithms discussed later in this chapter. Evolutionary algorithms are also reasonably well known as an approach for generating rule-based classification models, as will be briefly discussed in Section 4.3. Finally, a few attempts to extract classification rules from black-box models, such as artificial neural networks and support vector machines, were already made [44, 58], but this topic is out of the scope of this book.

2.2.1 On Predictive Accuracy

We mentioned earlier that, in order to evaluate the predictive accuracy of a classification model, we have to compute its predictive accuracy on a test set whose examples were not used during the training of the classification algorithm (i.e., during the construction of the classification model). We emphasize that the challenge

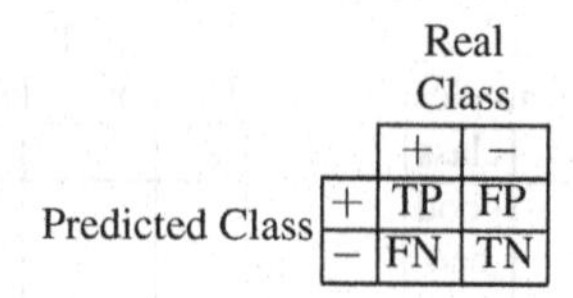

Fig. 2.2 Structure of a confusion matrix for a two-class problem

is really to achieve a high predictive accuracy in the test set, since achieving 100% classification accuracy in the training set can be considered a trivial task. In the latter case, the algorithm just needs to "memorize" the training data, i.e., to memorize what the class of each training example is. Hence, when asked to classify any training example, the algorithm would trivially do a kind of lookup in a table to find the actual class of that example. Although this very naive procedure would achieve classification accuracy of 100% in the training set, it involves no generalization at all. It is an extreme form of overfitting a classification model to the training data.

The majority of predictive accuracy measures are calculated using the elements of a confusion matrix [78]. A confusion matrix is an n x n matrix, where n is the number of classes in the problem at hand, that holds information about the correct and incorrect classifications made by the classification model. Figure 2.2 shows a confusion matrix for a two-class problem. As observed, the cells in the matrix show the number of examples correctly or incorrectly classified per class. The true positives (TP) and true negatives (TN) represent the number of examples correctly classified in the positive and negative classes, respectively. The false positives (FP) and false negatives (FN), in turn, represent the number of examples incorrectly classified as positive and negative, respectively. A simple measure of predictive accuracy (sometimes called the standard classification accuracy rate) is the number of correctly classified test examples ($TP+TN$) divided by the total number ($TP+FP+TN+FN$) of test examples. Although this measure is widely used, it has some disadvantages [41]. In particular, it does not take into account the problem of unbalanced class distribution.

For instance, suppose that a given dataset has two classes, where 99% of the examples belong to class c_1 and 1% belong to class c_2. If we measure predictive accuracy by the standard classification accuracy rate, the algorithm could trivially obtain a 99% predictive accuracy on the test set by classifying all test examples into the majority class in the training set, i.e., the class with 99% of relative frequency. This does not mean that the algorithm would be good at class predictions. It means rather that the standard classification accuracy rate is too easy to be maximized when the class distribution is very unbalanced, and so a more challenging measure of predictive accuracy should be used instead in such cases.

Other popular measures to evaluate predictive accuracy are sensitivity and specificity and some metric based on Receiver Operating Characteristic (ROC) analysis. The sensitivity (also named true positive rate) is calculated as the number of test examples correctly classified in the positive class (TP) divided by the total number of positive examples present in the test set ($TP+FN$). The specificity (or true negative rate), in contrast, is the total number of test examples correctly classified in the negative class (TN) divided by the total number of negative examples present in the test set ($TN+FP$). As sensitivity and specificity take into account the predictions for each class separately, they are preferred over the standard classification accuracy rate in problems with very unbalanced classes. The product of sensitivity and specificity, in particular, corresponds to a commonly used predictive accuracy measure.

The ROC analysis, in turn, plots a curve representing the true positive rate (sensitivity) against the false positive rate (1 − sensitivity) for a range of different values of a parameter of the classification model. Metrics such as the area under the curve (AUC) can then be used to assess predictive accuracy. Nonetheless, this approach is out of the scope of this book, and for more details the reader is referred to [29]. Moreover, good studies comparing different evaluation metrics and showing their advantages and disadvantages can be found in [14, 31].

At this point it is important to introduce a difference between academic or theoretical data mining research and data mining in practice, where the goal is to give useful knowledge to the user in real-world applications. In academic research the classification process usually involves two phases, using the training and test sets, respectively, as discussed earlier. However, in real-world data mining applications the classification process actually involves three phases, as follows.

First, in the training phase, the algorithm extracts some knowledge or classification model from the training set. Secondly, in the testing phase, one measures the predictive accuracy of that knowledge or model on the test set. The measured predictive accuracy is an estimate of the true predictive accuracy of the classification model over the entire unknown distribution of examples. In the third phase (which is normally absent in academic research), which will be called the "real-world application" phase, the classification model will be used to classify truly new, unknown-class examples, i.e., examples that were not available in the original dataset and whose class is truly unknown to the user. In practice, it is the predictive accuracy on this third dataset, to be available only in the future, that will determine the extent to which the classification algorithm was successful in practice. The predictive accuracy on the test set is very interesting for academic researchers, but it is less useful to the user in practice, simply because the classes of examples in the test set are known to the user. Those classes are simply hidden from the classification algorithm during its testing in order to simulate the scenario of future application of the classification model to new real-world data whose class is unknown by the user. Hence, it is hoped that the predictive accuracy on the test set is a good estimate of the predictive accuracy that will be achieved in future data, and this hope is justified by the assumption that the probability distribution of the future data is the same as the probability distribution of the test data. Note, however, that the extent to which this assumption is true cannot be measured before the future data is available.

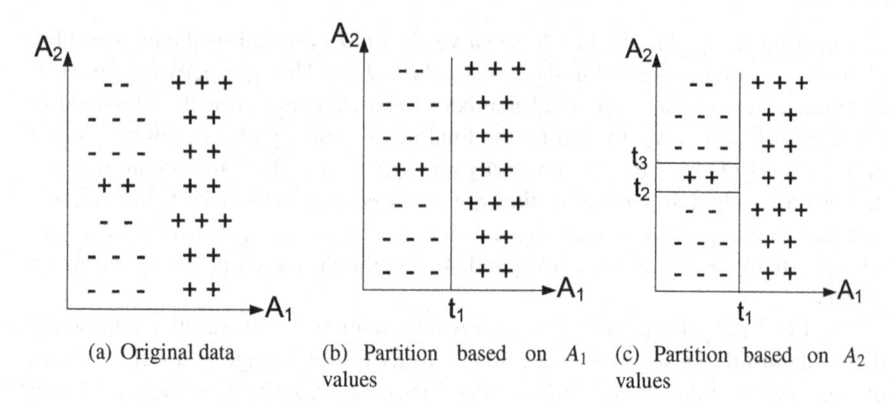

(a) Original data (b) Partition based on A_1 values (c) Partition based on A_2 values

Fig. 2.3 A geometric view of classification as a class separation problem

2.2.2 On Overfitting and Underfitting

Overfitting and underfitting are important problems in the classification task of data mining, and a well-designed classification algorithm should include procedures to try to mitigate these problems. The term *overfitting* refers to the case where a classification model is adjusted to details of the training data so much that it is representing very specific idiosyncracies of that data, and so it is unlikely to generalize well to the test dataset (unseen during training). The complementary term *underfitting* refers to the case where a classification model is under-adjusted to details of the training data that represent predictive relationships in the data. These concepts can be understood via a simple example, illustrated in Fig. 2.3. In this figure each training example is represented by a point in the two-dimensional space formed by two arbitrary attributes A_1 and A_2, and each example is labeled by a $+$ or $-$ sign, indicating whether it belongs to a positive or negative class. In this graphical perspective, in the simple case of a two-dimensional data space the goal of a classification algorithm can be regarded as finding lines (or curves) that separate examples of one class from examples of the other class. More generally, in a data space with more dimensions, the classification algorithm has to find hyperplanes separating examples from different classes.

Figure 2.3(a) shows the original training set, without any line separating the classes. Figures 2.3(b) and 2.3(c) show two possible results of applying a classification algorithm to the data in Fig. 2.3(a). The partitioning scheme shown in Fig. 2.3(c) can be represented by a set of classification rules, shown in Fig. 2.4. Note that in Fig. 2.3(c) the algorithm achieved a pure separation between the classes, while in Fig. 2.3(b) the left partition is not completely pure, since two positive examples belong to the same partition as a much larger number of negative examples. Hence, the partitioning scheme in Fig. 2.3(c) has a classification accuracy of 100% on the training set, while the scheme in Fig. 2.3(b) has a somewhat lower accuracy on the training set.

```
IF (A1 > t1) THEN (class = "+")
IF (A1 ≤ t1) AND (t2 ≤ A2 ≤ t3) THEN (class = "+")
IF (A1 ≤ t1) AND (A2 < t2) THEN (class = "-")
IF (A1 ≤ t1) AND (A2 > t3) THEN (class = "-")
```

Fig. 2.4 Classification rules corresponding to the data partition of Figure 2.2(c)

The interesting question, however, is which of these two partitioning schemes will be more likely to lead to a higher classification accuracy on examples in the test set, unseen during training. Unfortunately it is very hard to answer this question based only on the training set, without having access to the test set. Consider the two positive-class examples in the middle of the left partition in Fig. 2.3(b). There are two cases to consider. On the one hand, it is quite possible that these two examples are in reality noisy data, produced by errors in collecting and/or storing data. If this is the case, Fig. 2.3(b) represents a partitioning scheme that is more likely to maximize classification accuracy on unseen test examples than the partitioning in Fig. 2.3(c). The reason is that the latter would be mistakenly creating a small partition that covers only two noisy training examples, in which case we would say the classification model is overfitting the training data. On the other hand, it is possible that those two examples are in reality true exceptions in the data, representing a valid (though rare) relationship between attributes in the training data that is likely to be true in the test set too. In this case Fig. 2.3(c) represents a partitioning scheme that is more likely to maximize classification accuracy on unseen test examples than the partitioning in Fig. 2.3(b), because the latter would be underfitting the training data.

2.2.3 On the Comprehensibility of Discovered Knowledge

Although many data mining research projects use a measure of classification algorithm performance based only on predictive accuracy, it is accepted by many researchers and practitioners that, in many application domains, the comprehensibility of the knowledge (or patterns) discovered by a classification algorithm is another important evaluation criterion.

The motivation for discovering knowledge that is not only accurate but also comprehensible to the user can be summarized as follows. First, understanding the predictions output by a classification model can improve the user's confidence in those predictions. Note that in many application domains data mining is or should be used for decision support, rather than for automated decision making. That is, the knowledge discovered by a data mining algorithm is used to support a decision that will be made by a human user. This point is particularly important in applications such as medicine (where human lives are at stake), and science in general, and can also be important in some business and industrial applications where users might not invest a large amount of money on a decision recommended by a computational system if they did not understand the reason for that recommendation [23]. The importance of improving the user's confidence in a system's predictions (by providing him or her with a comprehensible model explaining the system's predictions) should not be underestimated. For instance, when there was a major accident in the Three Mile Island nuclear power plant, the automated system recommended a shutdown, but the human operator did not implement the shutdown because he or she did not believe in the system's recommendation [42].

Another motivation for discovering comprehensible knowledge is that a comprehensible classification model can be used not just for predicting the classes of test examples, but also for giving the user new insights about the data and the application domain. For instance, in the field of bioinformatics – involving the use of mathematical and/or computational methods for the analysis of biological data – data mining algorithms have been successful in discovering comprehensible models that provided new evidence confirming or rejecting biological hypotheses or led biologists to formulate new biological hypotheses. Several examples of this can be found in [34, 46].

Another motivation (related to the previous one) for comprehensible classification models is that their interpretation by the user can lead to the interesting detection of errors in the model and/or the data, caused for instance by limited quantity and quality of training data and/or by the use of an algorithm unsuitable for the data being mined. This point is discussed in more detail in the context of bioinformatics in [34, 71].

Knowledge comprehensibility is a kind of subjective concept, since a classification model can be little comprehensible for one user but very comprehensible for another user. However, to avoid difficult subjective issues, the data mining literature often uses an objective measure of "simplicity," based on classification model size – in general the smaller the classification model, the better the model. In particular, when the classification model is expressed by *if-then* classification rules, rule set size is usually used as a proxy to comprehensibility: in general, the smaller a rule set – i.e., the smaller the number of rules and the number of conditions per rule – the simpler the rule set is deemed to be. This concept of rule comprehensibility is still popular in the literature, probably due to its simplicity and objectivity, since the system can easily count the number of rules and rule conditions without depending on a subjective evaluation of a rule or rule set by the user. However, rule length is clearly a far from perfect criterion for measuring rule comprehensibility [60], [33, pp. 13–14].

One of the main problems of relying on rule length alone to measure rule comprehensibility is that this criterion is purely syntactical, ignoring semantic and cognitive science issues. Intuitively, a good evaluation of rule comprehensibility should go beyond counting conditions and rules, and in principle it should also include more subjective human preferences – which would involve showing the discovered rules to the user and asking for his or her subjective evaluation of those rules.

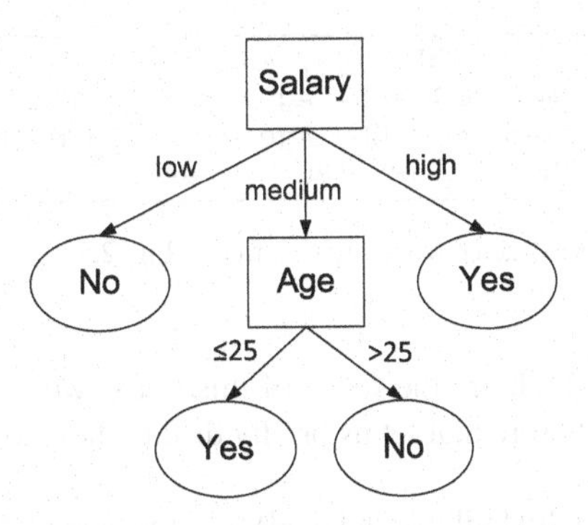

Fig. 2.5 A simple decision tree example

2.3 Decision Tree Induction

A decision tree is a graphical classification model in the form of a tree whose elements are essentially organized as follows: (a) every internal (non-leaf) node is labeled with the name of one of the predictor attributes; (b) the branches coming out from an internal node are labeled with values of the attribute labeling that node; and (c) every leaf node is labeled with a class. A very simple example of a decision tree is shown in Fig. 2.5, where the predictor attributes are *Salary* and *Age*, and the classes are *yes* and *no*, indicating whether or not a customer is expected to buy a given product.

In order to classify an example by using a decision tree, the example is pushed down the tree, following the branches whose attribute values match the example's attribute values until the example reaches a leaf node, whose class is then assigned to the example. For instance, considering the decision tree in Fig. 2.5, an example (customer) having the value *Salary* = *high* would be assigned class *yes*, regardless of the customer's *Age*, while an example having the values *Salary* = *medium* and *Age* = *32* would be assigned class *no*.

A decision tree can be straightforwardly converted into another kind of classification model, using a knowledge representation in the form of a set of *if-then* classification rules. This conversion can be done by creating one *if-then* rule for each path in the decision tree from the root node to a leaf node, so that there will be as many rules as leaf nodes. Each rule will have an antecedent (*if* part) consisting of all the attributes and corresponding values in the nodes and branches along that path, and will have a consequent (*then* part) predicting the class in the leaf node at the end of that path. For instance, the decision tree in Fig. 2.5 would be converted into the set of *if-then* classification rules shown in Fig. 2.6.

It should be noted, however, that a set of *if-then* classification rules can be induced from the data in a more direct manner using a rule induction algorithm, rather than by first inducing a decision tree and then converting it to a set of rules. Rule

```
IF (Salary = low) THEN (Buy = no)
IF (Salary = medium) AND (Age ≤ 25) THEN (Buy = yes)
IF (Salary = medium) AND (Age > 25) THEN (Buy = no)
IF (Salary = high) THEN (Buy = yes)
```

Fig. 2.6 Classification rules extracted from decision tree in Fig. 2.5

induction algorithms, which are the focus of this book, will be discussed in detail in the next section, but before that let us briefly discuss how to build a decision tree from the data.

A decision tree is usually built (or induced) by a top-down, "divide-and-conquer" algorithm [66], [11]. Other methods for building decision trees can be used. For instance, genetic programming can be used to build a decision tree, as will be discussed in Chapter 4.

Here we briefly describe the general top-down approach for decision tree building, which can be summarized by the pseudocode shown in Alg. 2.1 (adapted from [33]), where T denotes a set of training examples. Initially all examples of the training set are assigned to the root node of the tree. Then, unless a stopping criterion is satisfied, the algorithm selects a partitioning attribute and partitions the set of examples in the root node according to the values of the selected attribute. The goal of this process is to separate the classes so that examples of distinct classes tend to be assigned to different partitions. This process is recursively applied to the example subsets created by the partitions.

In the condition of the *if* statement, the most obvious stopping criterion is that the recursive-partitioning process must be stopped when all the data examples in the current training subset T have the same class, but other stopping criteria can be used [8, 66]. In the *then* statement, if all the examples in the just-created leaf node have the same class, the algorithm simply labels the node with that class. Otherwise, the algorithm usually labels the leaf node with the most frequent class occurring in that node. In the first step of the *else* statement the algorithm selects an attribute and a test over the values of that attribute in such a way that the outcomes of the selected test are useful for discriminating among examples of different classes [8, 11, 66]. In other words, for each test outcome, the set of examples that belong to the current tree node and satisfy that test outcome should be as pure as possible in the sense of having many examples of one class and few or no examples of the other classes.

The other steps of Alg. 2.1 are self-explanatory. In addition to the above basic decision tree building algorithm, one must also consider the issue of decision tree pruning. In a nutshell, pruning consists of simplifying (reducing the size of) a decision tree by removing irrelevant or unreliable leaf nodes or subtrees. The goal is to obtain a pruned tree smaller than and with a predictive accuracy at least as good as the original tree. Decision tree pruning is not further discussed here, and the reader is referred instead to [8, 12, 28, 55, 56, 64].

Algorithm 2.1: CreateDecisionTree

Initialize T with the set of all training examples
if current training set T satisfies a stopping criterion **then**
  Create a leaf node labeled with a class name and halt
else
  Select an attribute A to be used as a partitioning attribute and choose a test, over the values of A, with mutually exclusive and collectively exhaustive outcomes $O_1,...,O_k$
  Create a node labeled with the name of attribute A and create a branch, from the just-created node, for each test outcome
  Partition T into subsets $T_1,...,T_k$, such that each T_i, $i=1,...,k$, contains all examples in T with outcome O_i of the chosen test
  Apply this algorithm recursively to each training subset T_i, $i=1,...,k$

2.4 Rule Induction via the Sequential Covering Approach

The sequential covering strategy (also known as separate and conquer) is certainly the most explored and most used strategy to induce rules from data. It was first employed by the algorithms of the AQ family [53] in the late 1960s, and over the years was applied exaustively as the basic algorithm in rule induction systems.

In essence, the sequential covering strategy works as follows. It learns a rule from a training set (conquer step), removes from the training set the examples covered by the rule (separate step), and recursively learns another rule which covers the remaining examples. A rule is said to cover an example *e* when all the conditions in the antecedent of the rule are satisfied by the example *e*. For instance, the rule "*if (Salary > 150,000 euros) then Rich*" covers all the examples in the training set in which the value of salary is greater than 150,000 euros, regardless of the current value of the class attribute of an example.

The learning process goes on until a predefined criterion is satisfied. This criterion usually requires that all or almost all examples in the training set be covered by a rule in the classification model. Algorithm 2.2 shows the basic pseudocode for sequential covering algorithms producing unordered rule sets, where the discovered rules are applied to classify new test examples regardless of the order in which the rules were discovered. A somewhat different algorithm for generating ordered lists of rules will be presented in Section 2.4.1. Algorithm 2.3 describes the procedure *LearnOneRule* used by Alg. 2.2. Note that, in Alg. 2.2, the third step within the *while* loop calls Alg. 2.3 by passing, as a parameter, an initial rule R that has an empty rule antecedent (no conditions in its *if* part) and a consequent predicting the class C_i associated with the current iteration of the *for* loop. Algorithm 2.3 receives that initial rule and iteratively expands it, producing better and better rules until a stopping criterion is satisfied. The best rule built by Alg. 2.3 is then returned to Alg. 2.2, where it is added to the set of discovered rules. After that, the examples of class C_i covered by the just-discovered rule are removed from the training set, and so on, until rules for all classes have been discovered.

Algorithm 2.2: CreateRuleSet (produces unordered rules)

```
RuleSet = ∅
for each class C_i do
    Set training set T as the entire set of training examples
    while Stopping Criterion is not satisfied do
        Create an Initial Rule R
        Set class predicted by new rule R as C_i
        R'= LearnOneRule(R)
        Add R'to RuleSet
        Remove from T all class C_i examples covered by R'
Post-process RuleSet
return RuleSet
```

In Algs. 2.2 and 2.3, elements in italic represent a set of building blocks that can be instantiated to create different types of sequential covering algorithms. In the same manner, the block "*Create an Initial Rule R*" in Alg. 2.2 can be replaced by "*Create an empty Rule R*" (i.e., a rule with no conditions in its *if* part) or "*Create a Rule R using a random example from the training set.*" The block "*Evaluate CR*" in Alg. 2.3 could be replaced by "*Evaluate CR using accuracy*" or "*Evaluate CR using information gain.*"

Replacing building blocks in these basic algorithms by specific methods can create the majority of the existing sequential covering rule induction algorithms. This is possible because algorithms following the sequential covering approach usually differ from each other in four main ways: the representation of the candidate rules, the search mechanisms used to explore the space of the candidate rules, the way the candidate rules are evaluated, and the rule pruning method, although the last one can be absent [36, 78].

Before going into detail about these four ways, let us describe a couple of sequential covering algorithms that do not adopt exactly the pseudocode defined in Alg. 2.2. This is appropriate since it shows that attempts to improve this basic algorithm were made. Note that even though these new algorithms proved to be competitive with the traditional algorithms, currently the most used and accurate algorithms stick to the simple and basic approach described by the pseudocode in Alg. 2.2 (producing unordered rules) or by the pseudocode in Alg. 2.4 (producing ordered rules), which will be discussed later.

PNrule [1] and LERILS [18] are rule induction algorithms that slightly change the dynamics of the basic algorithm shown in Alg. 2.2. PN-Rules, for instance, is based on the concept that overfitting can be avoided by adjusting the trade-off between support (number of examples covered by a rule) and accuracy during the process of building a rule. The main difference between PN-Rules and the traditional algorithms is that the former finds two sets of rules: a set of P-rules and a set of N-rules. P-rules are learned first, favor high coverage, and are expected to cover most of the positive examples (i.e., examples belonging to the class predicted by the rule) in the training set. In a second step, a set of N-rules is created using only

Algorithm 2.3: LearnOneRule(R)

```
bestRule = R
candidateRules = ∅
candidateRules = candidateRules ∪ bestRule
while candidateRules ≠ ∅ do
    newCandidateRules= ∅
    for each candidateRule CR do
        Refine CR
        Evaluate CR
        if Refine Rule Stopping Criterion not satisfied then
            newCandidateRules = newCandidateRules ∪ CR
            if CR is better than bestRule then
                bestRule = CR
    candidateRules = Select b best rules in newCandidateRules
return bestRule
```

the set of examples covered by the P-rules. The idea is that the N-rules can exclude the false positives (examples covered by a P-rule but which belong to a different class from the predicted one) of the covered examples. This two-phase process is repeated for all the classes. During the classification of examples in the test set, a scoring method assesses the decisions of each binary classifier, and chooses the final classification.

Maybe it is not appropriate to call LERILS a sequential covering algorithm, since it does not remove examples from the training set after learning a rule. Instead, all the rules are learned using the whole set of training examples. However, apart from its not removing examples from the training set after creating a rule, all its other elements are based on conventional instances of the components found in the basic algorithm representing the sequential covering approach. LERILS also works in two phases. First it uses a bottom-up search combined with a random walk to produce a pool of k rules, where k is a parameter defined by the user (since the examples are not removed from the training set after creating a rule, a fixed number of rules is defined as the stopping criterion). In a second phase, it uses again a random procedure together with the minimum description length heuristic to combine the rules into a final rule set.

The literature in rule induction and especially sequential covering algorithms is very rich. There are several surveys, e.g., [36, 62], and papers comparing a variety of algorithms [4, 35, 54, 51]. The next subsections summarize the four main points to be considered when creating a sequential covering algorithm, and in particular the specific methods that can replace the building blocks presented in Algs. 2.2 and 2.3. For a more detailed description the reader is referred to the original papers describing the methods.

2.4.1 Representation of the Candidate Rules

The rule representation has a significant influence in the learning process, since some concepts can be easily expressed in one representation but hardly expressed in others. In particular, rules can be represented using propositional or first order logic.

Propositional rules are composed of selectors, which are associations between pairs of attribute-values, such as *Age* > *18*, *Salary* < *30,000*, or *Sex* = *male*. Besides the operators ">", "<", and "=", "≤", "≥", and "≠" are also available in some methods. The operators "=" and "≠" are used with nominal attributes, while the others are used with continuous attributes. The majority of the algorithms build a rule as a conjunction of selectors, although some algorithms also allow internal disjunctions (e.g., conditions like *Marital_Status* = *single* or *divorced*) and intervals (e.g., conditions like *30,000* ≤ *salary* ≤ *50,000*). The algorithms CN2 [20], C4.5Rules [66], and Ripper [22] are examples of propositional rule induction algorithms.

First-order logic rules are more sophisticated and have greater expressiveness power, since they can express relations between two attributes, generating rules with conditions such as *Income* > *Expenses*. When using a first-order representation, the concepts are usually represented as Prolog relations, like *Father(x,y)*. Methods that use this Prolog representation are classified as Inductive Logic Programming (ILP) systems [47, 48, 67].

ILP uses the same principles of rule induction algorithms, essentially replacing the concepts of conditions and rules by literals and clauses. In addition, ILP techniques allow the user to incorporate into the algorithm background knowledge about the problem, which helps to focus the search in promising areas of the search space. FOIL [65] and REP [13] use this representation.

Note, however, that ILP algorithms and other algorithms that discover first-order logic rules tend to be much more computationally expensive than their propositional counterparts. This is because the search space associated with first-order logic conditions is much larger than the search space associated with propositional logical conditions.

Apart from these two main representations, a few algorithms use some different representations. BEXA [72], for example, uses the multiple-valued extension to propositional logic to represent rules, while systems like FuzzConRI [79] use fuzzy logic. The latter, in special, are becoming more common.

Besides different rule representations, there are also different types of classification models that can be created when combining single rules into a collection of rules. The rule models generated by a rule induction algorithm can be ordered (also known as rule lists or decision lists) or unordered (rule sets).

In rule lists the order in which the rules are learned is important because rules will be applied in order when classifying new examples in the test set. In other words, the first rule in the ordered list that covers the new example will classify it, whereas subsequent rules will be ignored. In contrast, in rule sets the order in

Algorithm 2.4: CreateDecisionList (produces ordered rules)

```
RuleList = ∅
Set training set T as the entire set of training examples
repeat
    Create an Initial Rule R
    R' = LearnOneRule(R)
    Set consequent of learned rule R' as the most frequent class found in the set of
    examples covered by R'
    Add R' to RuleList
    Remove from T all the examples covered by R'
until Rules Stopping Criterion not satisfied
return RuleList
```

which the rules are produced is not important, since all the rules in the model are considered when classifying a new example. In the latter case, when more than one rule covers a new example, and the class predicted by them is not the same, a tiebreak criterion can be applied to decide which rule gives the best classification. Examples of these criteria are selecting the rule with higher coverage or higher value of heuristics like the Laplace estimation [24]. Algorithm 2.2 represents the algorithm employed to generate a set of unordered rules. A few changes have been introduced to it in order to generate an ordered decision list. Algorithm 2.4 describes the new pseudocode to generate a decision list following the sequential covering approach.

Comparing Algs. 2.2 and 2.4, the outer *for* loop from Alg. 2.2 is absent in Alg. 2.4 because decision list algorithms do not learn rules for each class in turn. Instead they try to learn, at each iteration, the best possible rule for the current training set, so that the class of each candidate rule will be chosen in a way to maximize the quality of that rule taking into account its antecedent. This is typically done by first building a candidate rule antecedent and then setting the candidate rule consequent to the most frequent class among all training examples covered by that candidate rule. Hence, rules predicting different classes can be learned in any order (the actual order depends on the training data), and the rules will be applied to classify test examples in the same order in which they were learned. The set of examples removed from the training set after a rule is learned also changes. While in Alg. 2.2 only the covered examples belonging to the current class C_i are removed, in Alg. 2.4 all the covered examples (no matter what their class) are removed. This happens because, as rule lists apply rules in order, if one rule covers a test example no other rules will have the chance to classify it.

Rule lists are, in general, considered more difficult to understand than rule sets. This is because in order to comprehend a given rule in an ordered list all the previous rules must also be taken into consideration [19]. Since the knowledge generated by rule induction algorithms is supposed to be analyzed and validated by an expert, rules at the end of the list become very difficult to understand, particularly in very long lists. Hence, unordered rules are often favored over ordered ones when comprehensibility is a particularly important rule evaluation criterion.

2.4.2 Search Mechanism

A rule induction algorithm acts like a search algorithm exploring a space of candidate rules. Its search mechanism has two components: a search strategy and a search method. The search strategy determines the region of the search space where the search starts and its direction, while the search method specifies which specializations or generalizations should be considered. The building blocks "*Create an Initial Rule R*" (in Alg. 2.2) and "*Refine CR*" (in Alg. 2.3) determine the search strategy of a sequential covering algorithm. "*Select b best rules in newCandidateRules*," in Alg. 2.3, implements its search method.

Broadly speaking, there are three kinds of search strategies: bottom-up, top-down, and bidirectional. A bottom-up strategy starts the search with a very specific rule, and iteratively generalizes it. This specific rule is usually an example from the training set, chosen at random (as in LERILS [18]) or using some heuristic. RISE [24], for example, starts considering all the examples in the training set as very specific rules. Then, for each rule, in turn, it searches for its nearest example of the same class according to a distance measure, and attempts to generalize the rule to cover that nearest example.

A top-down strategy, in contrast, starts the search with the most general rule and iteratively specializes it. The most general rule is the one that covers all examples in the training set (because it has an empty antecedent, which is always satisfied for any example). The top-down strategy is more frequently used by sequential covering algorithms but its main drawback is that as induction goes on, the amount of data available to evaluate a candidate rule decreases drastically, reducing the statistical reliability of the rules discovered. This usually leads to data overfitting and the small disjunct problem [15].

However, there are ways to prevent overfitting in top-down searches. A simple approach, with limited effectiveness, is to stop building rules when the number of examples in the current training set falls below some threshold. Note that this approach can miss some rare but potentially important rules representing potentially interesting exceptions to more general rules. Pruning methods, discussed in Section 2.4.4, are a more sophisticated approach to trying to avoid the problem of overfitting.

Finally, a bidirectional search is allowed to generalize or specialize the candidate rules. It is not a common approach, but it can be found in the SWAP-1 [77] and *Reconsider and Conquer* [7] algorithms. When looking for rules, SWAP-1 first tries to delete or replace the current rules' conditions before adding a new one. *Reconsider and Conquer*, in turn, starts the search with an empty rule, but after inserting the first best rule into the rule set, it backs up and carries on the search process using the candidate rules previously created.

After selecting the search strategy, a search method has to be implemented. The search method is a very important part of a rule induction algorithm since it determines which specializations or generalizations will be considered at each specialization or generalization step. Too many specializations or generalizations are not allowed due to computational time requirements, but too few may disregard good conditions and reduce the chances of finding a good rule.

The greedy and the beam searches are the most popular search methods. Greedy algorithms create an initial rule, specialize or generalize it, evaluate the extended rules created by the specialization or generalization operation, and keep just the best extended rule. This process is repeated until a stopping criterion is satisfied. PRISM [17] is just one among many algorithms that use greedy search.

Although they are fast and easy to implement, greedy algorithms have the well-known myopia problem: at each rule extension step, they make the best local choice, and cannot backtrack if later in the search the chosen path is not good enough to discriminate examples belonging to different classes. As a result, they do not cope well with attribute interaction [23, 32].

Beam search methods try to eliminate the drawbacks of greedy algorithms selecting, instead of one, the b best extended rules at each iteration, where b is the width of the beam. Hence, they explore a larger portion of the search space than greedy methods, coping better with attribute interaction. Nevertheless, learning problems involving very complex attribute interactions are still a very difficult problem for beam search algorithms, and for rule induction and decision tree algorithms in general [69]. In addition, beam search is of course more computationally expensive than greedy search. CN2, AQ and BEXA algorithms implement a beam search.

Note that if the parameter b is set to 1, we obtain a greedy search, so the greedy search can be considered a particular case of a beam search. However, in practice the term beam search is used only when b is set to a value greater than 1.

Apart from these two conventional search methods, some algorithms try to innovate in the search method in order to better explore the space of rules. ITRULE [70], for instance, uses a depth-first search, while LERILS [18] applies a randomized local search. Furthermore, stochastic search methods use evolutionary approaches, such as genetic algorithms and genetic programming, to explore the search space. Examples of systems using this approach will be discussed later.

In conclusion, the main problem with the search mechanism of sequential covering algorithms nowadays is that, regardless of a top-down or a bottom-up search, most of them use a greedy, hill-climbing procedure to look for rules.

A way to make these algorithms less greedy is to use an n-step look-ahead hill-climbing procedure. Hence, instead of adding or removing one attribute at a time from a rule, the algorithm would add or remove n conditions at a time. This approach was attempted by some decision tree induction algorithms in the past, but there is no strong evidence of whether look-ahead improves or harms the performance of the algorithm. While Dong and Kothari [25] concluded that look-ahead produces better decision trees (using a nonparametric look-ahead method), Murthy and Salzberg [57] argued it can produce larger and less accurate trees (using a one-step look-ahead method). A more recent study by Esmeir and Markovich [27] used look-ahead for decision tree induction, and found that look-ahead produces better trees and higher accuracies, as long as a large amount of time is available.

Look-ahead methods for rule induction were tested by Fürnkranz [37] in a bottom-up algorithm. One- and two- step look-aheads were used, and they sightly improved the accuracy of the algorithm in the datasets used in the experiments, but with the expense of quadratic time complexity. Nevertheless, further studies analyzing the impact of deeper look-ahead in the bottom-up and top-down approaches are needed to reach stronger conclusions about their effectiveness. It is believed that computational time requirement is one of the reasons that prevents the use of look-ahead in rule induction. However, in application domains in which time can be sacrificed for better classification models, it is an idea worth trying.

2.4.3 Rule Evaluation

The regions of the search space being explored by a rule induction algorithm can drastically change according to the rule evaluation heuristic used to assess a rule while it is being built. This section describes some of the heuristics used to estimate rule quality. In all the formulas presented, P represents the total number of positive examples in the training set, N represents the total number of negative examples in the training set, p represents the number of positive examples covered by a rule R, and n the number of negative examples covered by a rule R. In Alg. 2.3, the building block "*Evaluate CR*" is responsible for implementing rule evaluation heuristics.

When searching for rules, the first aim of most of the sequential covering algorithms is to find rules that maximize the number of positive covered examples and, at the same time, minimize the number of negative covered examples. It is important to note that these two objectives are conflicting because as the number of positive covered examples increases, the number of negative covered examples tends to also increase. Examples of rule evaluation heuristics used by these algorithms are confidence, Laplace estimation, M-estimate, and LS content.

Confidence (also known as precision or purity) is the simplest rule evaluation function and is described as in Eq. (2.1).

$$confidence(R) = \frac{p}{p+n} \tag{2.1}$$

It is used by SWAP-1, and its main drawback is that it is prone to overfitting. As an example of this problem, consider two rules: a rule R_1 covering 95 positive examples and five negative examples (confidence = 0.95), and a rule R_2 covering two positive examples and no negative examples (confidence = 1). An algorithm choosing a rule based on the confidence measure will prefer R_2. This is undesirable, because R_2 is not a statistically reliable rule, being based on such a small number of covered examples. In order to overcome this problem with the confidence measure, the Laplace estimation (or "correction") measure was introduced, and it is defined in Eq. (2.2).

In Eq. (2.2), *nClasses* is the number of classes available in the training set. Using this heuristic, rules with apparently high confidence but very small statistical support are penalized. Consider the previously mentioned rules R_1 and R_2 in a two-class problem. The Laplace estimation values are 0.94 for R_1 and 0.75 for R_2. R_1 would be preferred over R_2, as it should be. Laplace estimation is used by the CN2 [19] and BEXA [72] algorithms.

Note that a better name for the above Laplace estimation would probably be the confidence measure with Laplace correction, because the term Laplace estimation is very generic, and some kind of Laplace correction could in principle be applied to measures of rule quality different from the confidence measure. In any case, in this book we will use the term Laplace estimation for short, to be consistent with most of the literature in this area.

$$laplaceEstim(R) = \frac{p+1}{p+n+nClasses} \tag{2.2}$$

M-estimate [26] is a generalization of the Laplace estimation, where a rule with 0 coverage is evaluated considering the classes' a priori probabilities, instead of $1\,/\,nClasses$. More precisely, M-estimate is computed by Eq. (2.3), where M is a parameter. Equation (2.3) corresponds to adding M virtual examples to the current training set, distributed according to the prior probabilities of the classes. Hence, higher values of M give more importance to the prior probabilities of classes, and their use is appropriate in datasets with high levels of noise.

$$M\text{-}estimate(R) = \frac{p+m\frac{P}{P+N}}{p+n+m} \tag{2.3}$$

The LS content measure, shown in Eq. (2.4), divides the proportion of positive examples covered by the rule by the proportion of negative examples covered by the rule, both proportions estimated using the Laplace estimation. $P + nClasses$ and $N + nClasses$ can be omitted because they are constant during the rule refinement process. The LS content is used by the HYDRA [2] algorithm.

$$LS\,content(R) = \frac{\frac{p+1}{P+nClasses}}{\frac{n+1}{N+nClasses}} \cong \frac{p+1}{n+1} \tag{2.4}$$

From these four described heuristics, the Laplace estimation and the m-estimate seem the most successfully used, mainly because of their small sensitivity to noise.

A second desired feature in rules is rule simplicity. Rule size is the most straightforward measure of simplicity, but more complicated heuristics such as the minimum description length [65] can also be applied. Nevertheless, heuristics to measure simplicity are most of the time combined with other rule evaluation criteria.

The minimum description length provides a trade-off between size and accuracy using concepts of information theory. It first estimates the size of a "theory" (a rule set in the context of rule induction algorithms), measured in number of bits, and then adds to it the number of bits necessary to encode the exceptions relative to the theory, using an information loss function [78]. The aim is to minimize the total description length of the theory and its exceptions.

Within the group of rule evaluation heuristics that measure the coverage/size of a rule are the ones based on gain, which compute the difference in the value of some heuristic function measured between the current rule and its predecessor. Information gain is the most popular of these heuristics, and it is defined as in Eq. (2.5), where *R'* represents a specialization or generalization of the rule *R*. In Eq. (2.5), the logarithm of the rule confidence is also known as the information content of the rule, and it can also be used as a heuristic function by itself. The information gain measure is used by the PRISM [17] and Ripper [22] algorithms.

$$infoGain(R) = -log(confidence(R)) - log(confidence(R')) \tag{2.5}$$

In [38, 39], Fürnkranz and Flach analyzed the effect of commonly used rule evaluation heuristic functions, and found that many of them are equivalent. They concluded that there are two basic prototypes of heuristics, namely precision and the cost-weighted difference between positive and negative examples. Precision is equivalent to the confidence measure described in Eq. (2.1), and the cost-weighted difference is defined in Eq. (2.6), where *d* is an arbitrary cost.

$$costWeigth = p - dn \tag{2.6}$$

They also interpreted the Laplace estimation and the M-estimate as a trade-off between these two basic heuristics, and again recognized their success due to their smaller sensitivity to noise.

There is a last property desired in discovered rules but not often considered by rule induction algorithms: interestingness, in the sense that ideally a rule should also be novel (or surprising) and potentially useful (actionable) to the user. Rule interestingness is very difficult to measure, but as shown by Tsumoto [73], it is very desirable in practice when the rules will be analyzed by the user. He demonstrated that from 29,050 rules found by a rule induction algorithm only 220 (less than 1%) were considered interesting by a user. When measuring rule interestingness, two approaches can be followed: a user-driven approach or a data-driven approach.

The user-driven approach uses the user's background knowledge about the problem to evaluate rules. For instance, Liu et al. [52] and Romao et al. [68] used, as background knowledge, the general impressions of users about the application domain in the form of *if-then* rules. The general impressions were matched with the discovered rules in order to find, for example, rules with the same antecedent and different consequents from the general impressions, and, therefore, surprising rules (in the sense that they contradict some general impressions of the user).

In contrast, data-driven approaches measure interestingness based on statistical properties of the rules, in principle without using the user's background knowledge. A review of data-driven approaches can be found in [43]. Measuring the interestingness of rules to the user in an effective way without the need for the user background knowledge may sound appealing at first glance. Nonetheless, the effectiveness of this approach is questionable, given the great variability of the backgrounds and interests of users. It is important to note that a couple of studies tried to calculate the correlation between the value of these data-driven rule interestingness measures and the real subjective interest of users on the rules, and those studies suggest this correlation is relatively low [16, 59]. Among data-driven measures, measures that favor interesting "exception rules," i.e., rules representing interesting exceptions to general patterns represented by more generic rules, seem more promising, at least in the sense that exception rules tend to be more likely to represent novel (previously unknown) knowledge to users than very generic rules.

At last, it is interesting to point out that, intuitively, complete and incomplete rules should be evaluated using different heuristics. The reason is that, while in incomplete rules, i.e., rules that are still being constructed, there is a strong need to cover as many positive examples as possible, a major goal of complete rules is also to cover as few negative examples as possible. Most algorithms use the same heuristic to evaluate both complete and incomplete rules. It would be interesting to evaluate the effects of using different measures to evaluate rules in different stages of the specialization or generalization processes.

2.4.4 Rule Pruning Methods

The first algorithms developed using the sequential covering approach searched the data for complete and consistent rule sets. This means they were looking for rules that covered all the examples in the training set (complete) and that covered no negative examples (consistent). However, real-world datasets are rarely complete and usually noisy.

Pruning methods were introduced to sequential covering algorithms to avoid overfitting and to handle noisy data, and are divided in two categories: pre- and post-pruning. Pre-pruning methods stop the refinement of the rules before they become too specific or overfit the data, while post-pruning methods find a complete and consistent rule or rule set, and later try to simplify it. Pre-pruning methods are implemented through the building blocks "*Stopping Criterion*" in Alg. 2.2 and "*Refine Rule Stopping Criterion*" in Alg. 2.3. Post-pruning uses the building block "*Post-process RuleSet*" in Alg. 2.2.

Pre-pruning methods include stopping a rule's refinement process when some predefined condition is satisfied, allowing it to cover a few negative examples. They also apply the same sort of predefined criterion to stop adding rules to the classification model, leaving some examples in the training set uncovered.

Along with the most common criteria applied for pre-pruning are the use of a statistical significance test (used by CN2); requiring a rule to have a minimum accuracy (or confidence), such as in IREP [40], where rule accuracy has to be at least equal to the accuracy of the empty rule; or associating a cutoff stopping criterion to some other heuristics.

The statistical significance test used by CN2 compares the observed class distribution among examples satisfying the rule with the expected distribution that would result if the rule had selected examples randomly. It provides a measure of distance between these two distributions. The smaller the distance, the more likely that the distribution created by the rule is due to chance.

Post-pruning methods aim to improve the learned classification model (rule or rule set) after it has been built. They work by removing rules or rules' conditions from the model, while trying to preserve or improve the classification accuracy in the training set. Among the most well-known post-pruning techniques are reduced error pruning (REP) [13] and GROW [21]. These two techniques follow the same principles. They divide the training data into two parts (grow and prune sets), learn a model using the grow set, and then prune it using the prune set. REP prunes rules using a bottom-up approach while GROW uses a top-down approach. Hence, instead of removing the worst condition from the rules while the accuracy of the model remains unchanged (as REP does), GROW adds to a new empty model the best generalization of the current rules.

When comparing pre- and post-pruning methods, each of them has its advantages and pitfalls. Though pre-pruning methods are faster, post-pruning methods usually produce simpler and more accurate models (at the cost of efficiency, since some rules are learned and then simply discarded from the model). Intuitively, this is due to the fact that post-pruning has more information (the complete learned classification model) available to make decisions, and so it tends to be less "shortsighted" than pre-pruning. However, methods which learn very specific rules and later prune them in a different set of data often have a problem related to the size of the data. If the amount of training data is limited, dividing it in two subsets can have a negative effect since the rule induction algorithm may not get statistical support from the data when finding or pruning rules.

In any case, pruning complete rule sets is not as straightforward as pruning decision trees. Considering that most of the rule pruning literature was borrowed from the tree pruning literature [66], it is necessary to keep in mind that pruning a subtree always maintains full coverage of the dataset, while pruning rules can leave currently covered examples uncovered, and the algorithm may have no resources to reverse this situation.

In an attempt to solve the problems caused by pre- and post-pruning techniques, some methods combine or integrate them to get the best of both worlds. Cohen [21], for example, combined the minimum description length criterion (to produce an initially simpler model) with the GROW algorithm.

I-REP [40] and its improved version Ripper [22] are good examples of integrated approaches. Their rule pruning techniques follow the same principles of REP [13], but they prune each rule right after it is created, instead of waiting for the complete model to be generated. After one rule is produced, the covered examples are excluded from both the grow and prune sets, and the remaining examples are redivided into two new grow and prune sets.

The main differences between I-REP and Ripper lie in Ripper's optimization process, which is absent in I-REP, and on the heuristics used for pruning rules and stopping the addition of rules to the rule set. The optimization process considers each rule in the current rule set in turn, and creates two alternative rules from them: a replacement rule and a revision rule [22]. After that, a decision is made on whether the model should keep the original rule, the replacement, or the revision rule, based on the minimum description length criterion.

If at the end of the optimization process there are still positive examples in the training set that are not covered by any of the rules, the algorithm can be applied again to find new rules which will cover the remaining uncovered positive examples.

2.5 Meta-learning

It is well known that no classification algorithm is the best across all applications domains and datasets, and that different classification algorithms have different inductive biases that are suitable for different application domains and datasets. Hence, meta-learning has emerged as an approach to automatically find out the best classification algorithm for a given target dataset.

Meta-learning, as the term suggests, involves learning about the results of (base-level) learning algorithms. The term *meta* has the same kind of meaning here as in other computer science areas; e.g., in the area of databases it is common to say that meta-data describes the base-level data stored in a database system. Since this book focuses on the classification task, let us study meta-learning in this context only, though meta-learning can of course be applied to other data mining tasks. Even in the context of classification only, there are several different types of meta-learning. This section will review just two major types of meta-learning, namely meta-learning for classification algorithm selection and stacked generalization. For a review of other types of meta-learning, see [9, 74, 75].

2.5.1 Meta-learning for Classification Algorithm Selection

As discussed earlier, a classification algorithm learns from a given dataset in a specific application domain, i.e., it learns from a given training set a classification model that is used to predict the classes of examples in the test set. In this context, a major type of meta-learning involves learning, from a number of datasets (from different application domains), relationships between datasets and classification algorithms. The basic idea is to create a meta-dataset where each meta-example represents an entire dataset. Each meta-example is described by a set of meta-attributes, each of which represents a characteristic of the corresponding dataset. Each meta-example is associated with a meta-class. This is often the name of the *classification algorithm* that is recommended as the best algorithm for the corresponding dataset out of a set of candidate classification algorithms. Hence, the set of meta-classes is the set of candidate classification algorithms. Once such a meta-dataset has been created, a meta-classification algorithm is used to learn a meta-classification model that predicts the best classification algorithm (meta-class) for a given dataset (meta-example) based on characteristics describing that dataset (its meta-attributes). In most meta-learning research, by "best" classification algorithm is meant the algorithm with the highest predictive accuracy on a given dataset, but it is certainly possible to consider other performance criteria too. For instance, one can measure an algorithm's performance by a combination of its predictive accuracy and its computational time [3, 10].

Note that the meta-classification algorithm can be actually any conventional classification algorithm – common choices are a nearest neighbor (instance-based learning) algorithm [10] or a decision tree induction algorithm [45]. It is called a meta-classification (or meta-learning) algorithm because it is applied to data at a meta-level, where each meta-example consists of meta-attributes describing characteristics of an entire dataset and each meta-class is the name of the classification algorithm recommended for that dataset. Hence, the crucial issue is how to construct the meta-dataset.

Like any other dataset for classification, the meta-dataset is divided into a meta-training set and a meta-test set. In order to assign meta-classes to the meta-training set, for each meta-example in the meta-training set, the system applies each of the candidate classification algorithms to the dataset corresponding to that meta-example, and measures their predictive accuracy according to some method (e.g., cross-validation [78]). Then, the most accurate classification algorithm for that dataset is chosen as the meta-class for that meta-example. Note that this approach for generating the meta-classes of the meta-examples is computationally expensive, since it involves running a number of different classification algorithms on each dataset in the meta-training set. One approach to significantly reduce processing time is to actively select only a subset of the available datasets to have meta-classes generated by the above procedure, i.e., to perform some form of meta-example selection to identify the most relevant meta-examples for meta-learning [63].

A crucial and difficult research problem consists of deciding which meta-attributes will be used to describe the datasets represented by the meta-examples. It is far from trivial to choose a set of meta-attributes that captures predictive information about which classification algorithm will be the best for each dataset. Broadly speaking, three types of meta-attributes are often found in the literature: (a) statistical and information-theoretic meta-attributes; (b) classification model-based meta-attributes; and (c) landmarking-based meta-attributes. Let us briefly review each of those types of meta-attributes.

Statistical and information-theoretic meta-attributes can capture a wide range of dataset characteristics, varying from very simple measures such as number of examples, number of nominal attributes, and number of continuous attributes in the dataset, to measures like the average value of some measure of correlation between each attribute and the class attribute (e.g., average information gain of the nominal attributes). This type of meta-attribute was extensively used in the well-known Statlog project [54]. One drawback of some meta-attributes used in that project is that their calculation was computationally expensive, sometimes even slower than running some candidate classification algorithms, so that in principle it would be more effective to use that computational time to run a classification algorithm. In any case, the Statlog project produced several interesting results (for a comprehensive discussion of this project, see [54]) and was very influential in meta-learning research.

Another drawback of many statistical and information-theoretic meta-attributes typically used in meta-learning is that they capture just very high-level, coarse-grained properties of each dataset as a whole, rather than more detailed properties of (parts of) the dataset that are potentially useful to predict the performance of classification algorithms [75]. Alternatively, one could have finer-grained meta-attributes representing some kind of probability distribution of some dataset characteristic. For instance, in [45] the percentage of missing values for each attribute is computed, and then, instead of creating a single meta-attribute representing the average percentage of missing values across all attributes, meta-attributes are created to represent a histogram of missing values. That is, the first meta-attribute represents the proportion of attributes having between 0% and 10% of missing values, the second meta-attribute represents the proportion of attributes having between 10% and 20% of missing values, and so on. These finer-grained meta-attributes could lead to a better characterization of datasets at the price of a significant increase in the number of meta-attributes.

Classification model-based meta-attributes are produced for each meta-example in two phases. First, the system runs a classification algorithm on the dataset represented by the meta-example and builds a classification model for that dataset. Secondly, some characteristics of that model are automatically extracted and used as values of meta-attributes for that meta-example. Intuitively, the classification algorithm should ideally be relatively fast and produce a kind of understandable model that facilitates the design and interpretation of the meta-attributes by the researcher or the user. A typical choice of classification algorithm in this context is a decision tree induction algorithm, since it is relatively fast and produces models in an understandable knowledge representation, viz. decision trees. For instance, from a decision tree one can extract meta-attributes such as the ratio of the number of tree nodes to the number of attributes in the dataset [5], which can indicate the number of irrelevant attributes in the dataset, particularly for datasets with a large number of attributes.

Landmarking-based meta-attributes are based on the idea of using a set of relatively simple (and fast) classification algorithms as "landmarkers" for predicting the performance of more sophisticated (and typically much slower) classification algorithms [50, 61, 74]. Such meta-attributes are produced for each meta-example as follows. The system runs each of the landmarker algorithms on the dataset represented by the meta-example. Some measure of predictive accuracy for each landmarker is then used as the value of a meta-attribute, so that this approach produces as many meta-attributes as the number of landmarkers. This approach requires a diverse set of landmarkers to be effective, i.e., different landmarkers should measure different dataset properties [61]. In addition, the landmarker algorithms' predictive accuracy should be correlated with the predictive accuracy of the candidate algorithm being landmarked [50], since one wants to use the performance of the former to predict the performance of the latter. Instead of using simplified algorithms as landmarkers, some variations of the landmarking approach use other kinds of landmarkers, such as simplified versions of the data (sometimes called data sampling landmarking) and learning curves containing information about an algorithm's predictive accuracy for a range of data sample sizes [49].

2.5.2 *Stacked Generalization: Meta-learning via a Combination of Base Learners' Predictions*

In this approach a meta-classification (meta-learning) algorithm learns from the predictions of base classification algorithms (base learners) [9]. In the training phase, each base classification algorithm learns, from a given training set, a base classification model. The class predicted by each base classification model for a given example is used as a meta-attribute, so this approach produces n meta-attributes, where n is the number of base classification algorithms. Once all n base classification algorithms have been trained and the values of all n meta-attributes have been computed for all training examples, we have a meta-training set. This meta-training set can contain as attributes either just the n meta-attributes or both the n meta-attributes and the original base attributes of the training data. In any case, a meta-classification algorithm learns, from the meta-training set, a meta-classification model that predicts the class of a meta-example (i.e., an example containing the meta-attributes). In the testing phase, each test example is first classified by the base classification algorithms in order to produce its meta-attributes. Then the meta-classification model learned from the meta-training set is used to predict the class of that test meta-example.

It is important to note that this approach is very different from the meta-learning approach discussed in Section 2.5.1, because in the former the meta-learning algorithm uses information derived from a single dataset, while in the latter the meta-learning algorithm uses information derived from a number of datasets, learning a relationship between characteristics (meta-attributes) of different datasets and the performance of different classification algorithms. In addition, in stacked generalization each meta-example corresponds to an example in a given dataset (which is called meta-example because its attribute vector representation contains meta-attributes referring to classes predicted), and the output of the meta-classification algorithm is a predicted class for a meta-example. In contrast, in the meta-learning approach discussed in Section 2.5.1 each meta-example corresponds to an entire dataset, and the output of the meta-classification algorithm is the name of an algorithm that is recommended as the best classification algorithm that should be applied to the dataset corresponding to the meta-example.

2.6 Summary

This chapter introduced the basic concepts of the data mining classification task. In particular, it discussed how to evaluate the classification models produced and avoid overfitting and underfitting. It also stressed the importance of human-comprehensible knowledge, and reviewed the two most used types of methods to induce classification rules: decision trees and sequential covering algorithms. This latter type of classification algorithm, in particular, is the focus of this book.

While describing sequential covering rule induction algorithms, we identified their four main components: rule representation, search mechanism, rule evaluation, and pruning. We also showed how each of these four components could be implemented. At last, we discussed two major types of meta-learning methods, namely classification algorithm selection and stacked generalization.

The genetic programming (GP) system proposed in this book to automate the design of rule induction algorithms (to be described in detail in Chapter 5) can be regarded as a different type of meta-learning, namely "constructive meta-learning," since the GP system will construct a new classification algorithm.

References

1. Agarwal, R., Joshi, M.V.: PNrule: a new framework for learning classifier models in data mining. In: Proc. of the 1st SIAM Int. Conf. in Data Mining, pp. 1–17 (2001)
2. Ali, K.M., Pazzani, M.J.: Hydra: a noise-tolerant relational concept learning algorithm. In: R. Bajcsy (ed.) Proc. of the 13th Int. Joint Conf. on Artificial Intelligence (IJCAI-93), pp. 1064–1071 (1993)
3. Ali, S., Smith, K.: On learning algorithm selection for classification. Applied Soft Computing **6**, 119–138 (2006)
4. An, A., Cercone, N.: Rule quality measures for rule induction systems: Description and evaluation. Computational Intelligence **17**(3), 409–424 (2001)
5. Bensusan, H., Giraud-Carrier, G., Kennedy, C.: A higher-order approach to meta-learning. In: Proc. of the Workshop on Meta-Learning (ECML-00), pp. 109–118 (2000)
6. Berthold, M., Hand, D.J. (eds.): Intelligent Data Analysis: An Introduction. Springer-Verlag New York, Secaucus, NJ, USA (1999)
7. Boström, H., Asker, L.: Combining divide-and-conquer and separate-and-conquer for efficient and effective rule induction. In: S. Džeroski, P. Flach (eds.) Proc. of the 9th Int. Workshop on Inductive Logic Programming (ILP-99), pp. 33–43 (1999)
8. Bramer, M.: Principles of Data Mining. Springer (2007)
9. Brazdil, P., Giraud-Carrier, C., Soares, C., Vilalta, R.: Metalearning: applications to data mining. Springer (2009)
10. Brazdil, P., Soares, C., da Costa, J.: Ranking learning algorithms: using IBL and meta-learning on accuracy and time results. Machine Learning **50**, 251–277 (2003)
11. Breiman, L., Friedman, J., Olshen, R., Stone, C.: Classification and Regression Trees. Wadsworth (1984)
12. Breslow, L., Aha, D.: Simplifying decision trees: a survey. The Knowledge Engineering Review **12**(1), 1–40 (1997)
13. Brunk, C.A., Pazzani, M.J.: An investigation of noise-tolerant relational concept learning algorithms. In: L. Birnbaum, G. Collins (eds.) Proc. of the 8th Int. Workshop on Machine Learning, pp. 389–393. Morgan Kaufmann (1991)
14. Caruana, R., Niculescu-Mizil, A.: Data mining in metric space: an empirical analysis of supervised learning performance criteria. In: Proc. of the 10th Int. Conf. on Knowledge Discovery and Data Mining (KDD-04), pp. 69–78. ACM Press (2004)
15. Carvalho, D.R., Freitas, A.A.: A hybrid decision tree/genetic algorithm for coping with the problem of small disjuncts in data mining. In: D. Whitley, D. Goldberg, E. Cantu-Paz,

L. Spector, I. Parmee, H. Beyer (eds.) Proc. of the Genetic and Evolutionary Computation Conf. (GECCO-00), pp. 1061–1068. Morgan Kaufmann, Las Vegas, Nevada, USA (2000)
16. Carvalho, D.R., Freitas, A.A., Ebecken, N.: Evaluating the correlation between objective rule interestingness measures and real human interest. In: A. Jorge, L. Torgo, P. Brazdil, R. Camacho, J. Gama (eds.) Proc. of the 9th European Conf. on Principles and Practice of Knowledge Discovery in Databases (PKDD-05), pp. 453–461. Springer Verlag (2005)
17. Cendrowska, J.: Prism: an algorithm for inducing modular rules. International Journal of Man-Machine Studies **27**, 349–370 (1987)
18. Chisholm, M., Tadepalli, P.: Learning decision rules by randomized iterative local search. In: L. Birnbaum, G. Collins (eds.) Proc. of the 19th Int. Conf. on Machine Learning (ICML-02), pp. 75–82. Morgan Kaufmann (2002)
19. Clark, P., Boswell, R.: Rule induction with CN2: some recent improvements. In: Y. Kodratoff (ed.) Proc. of the European Working Session on Learning on Machine Learning (EWSL-91), pp. 151–163. Springer-Verlag, New York, NY, USA (1991)
20. Clark, P., Niblett, T.: The CN2 induction algorithm. Machine Learning **3**, 261–283 (1989)
21. Cohen, W.W.: Efficient pruning methods for separate-and-conquer rule learning systems. In: Proc. of the 13th Int. Joint Conf. on Artificial Intelligence (IJCAI-93), pp. 988–994. France (1993)
22. Cohen, W.W.: Fast effective rule induction. In: A. Prieditis, S. Russell (eds.) Proc. of the 12th Int. Conf. on Machine Learning (ICML-95), pp. 115–123. Morgan Kaufmann, Tahoe City, CA (1995)
23. Dhar, V., Chou, D., Provost, F.J.: Discovering interesting patterns for investment decision making with GLOWER – a genetic learner overlaid with entropy reduction. Data Mining and Knowledge Discovery **4**(4), 251–280 (2000)
24. Domingos, P.: Rule induction and instance-based learning: a unified approach. In: Proc. of the 14th Int. Joint Conf. on Artificial Intelligence (IJCAI-95), pp. 1226–1232 (1995)
25. Dong, M., Kothari, R.: Look-ahead based fuzzy decision tree induction. IEEE Transactions on Fuzzy Systems **9**(3), 461–468 (2001)
26. Dzeroski, S., Cestnik, B., Petrovski, I.: Using the m-estimate in rule induction. Journal of Computing and Information Technology **1**(1), 37–46 (1993)
27. Esmeir, S., Markovitch, S.: Lookahead-based algorithms for anytime induction of decision trees. In: Proc. of the 21th Int. Conf. on Machine Learning (ICML-04) (2004)
28. Esposito, F., Malerba, D., Semeraro, G.: Decision tree pruning as search in the state space. In: Proc. of the European Conf. on Machine Learning (ECML-93), pp. 165–184. Springer (1993)
29. Fawcett, T.: ROC graphs: notes and practical considerations for data mining researchers. Tech. Rep. HPL-2003-4, HP Labs (2003)
30. Fayyad, U.M., Piatetsky-Shapiro, G., Smyth, P.: From data mining to knowledge discovery: an overview. In: U.M. Fayyad, G. Piatetsky-Shapiro, P. Smyth, R. Uthurusamy (eds.) Advances in Knowledge Discovery and Data Mining. AAAI/MIT Press (1996)
31. Flach, P.: The geometry of ROC space: understanding machine learning metrics through ROC isometrics. In: Proc. 20th Int. Conf. on Machine Learning (ICML-03), pp. 194–201. AAAI Press (2003)
32. Freitas, A.A.: Data Mining and Knowledge Discovery with Evolutionary Algorithms. Springer-Verlag (2002)
33. Freitas, A.A., Lavington, S.H.: Mining Very Large Databases with Parallel Processing. Kluwer Academic Publishers (1998)
34. Freitas, A.A., Wieser, D., Apweiler, R.: On the importance of comprehensible classification models for protein function prediction. IEEE/ACM Transactions on Computational Biology and Bioinformatics (in press)
35. Fürnkranz, J.: Pruning algorithms for rule learning. Machine Learning **27**(2), 139–171 (1997)
36. Fürnkranz, J.: Separate-and-conquer rule learning. Artificial Intelligence Review **13**(1), 3–54 (1999)
37. Fürnkranz, J.: A pathology of bottom-up hill-climbing in inductive rule learning. In: Proc. of the 13th Int. Conf. on Algorithmic Learning Theory (ALT-02), pp. 263–277. Springer-Verlag, London, UK (2002)

38. Fürnkranz, J., Flach, P.: An analysis of rule evaluation metrics. In: Proc. 20th Int. Conf. on Machine Learning (ICML-03), pp. 202–209. AAAI Press (2003)
39. Fürnkranz, J., Flach, P.A.: ROC 'n' rule learning: towards a better understanding of covering algorithms. Machine Learning **58**(1), 39–77 (2005)
40. Fürnkranz, J., Widmer, G.: Incremental reduced error pruning. In: Proc. the 11th Int. Conf. on Machine Learning (ICML-94), pp. 70–77. New Brunswick, NJ (1994)
41. Hand, D.J.: Construction and Assessment of Classification Rules. Wiley (1997)
42. Henery, R.: Classification. In: D. Michie, D. Spiegelhalter, C. Taylor (eds.) Machine Learning, Neural and Statistical Classification. Ellis Horwood (1994)
43. Hilderman, R.J., Hamilton, H.J.: Knowledge Discovery and Measures of Interest. Kluwer Academic Publishers, Norwell, MA, USA (2001)
44. Jacobsson, H.: Rule extraction from recurrent neural networks: A taxonomy and review. Neural Computation **17**, 1223–1263 (2005)
45. Kalousis, A., Hilario, M.: Model selection via meta-learning: a comparative study. In: Proc. of the 12th IEEE Int. Conf. on Tools with Artificial Intelligence (ICTAI-00), pp. 406–413 (2000)
46. Karwath, A., King, R.: Homology induction: the use of machine learning to improve sequence similarity searches. BMC Bioinformatics **3**(11), online publication (2002). http://www.pubmedcentral.nih.gov/articlerender.fcgi?artid=107726
47. Lavrac, N., Dzeroski, S.: Inductive Logic Programming: Techniques and Applications. Routledge, New York, (1993)
48. Lavrac, N., Dzeroski, S. (eds.): Relational Data Mining. Springer-Verlag, Berlin (2001)
49. Leite, R., Brazdil, P.: Predicting relative performance of classifiers from samples. In: Proc. of the 22nd Int. Conf. on Machine Learning (ICML-05), pp. 497–504 (2005)
50. Ler, D., Koprinska, I., Chawla, S.: Comparisons between heuristics based on correlativity and efficiency for landmarker generation. In: Proc. of the 4th Int. Conf. on Hybrid Intelligent Systems (HIS-04) (2004)
51. Lim, T., Loh, W., Shih, Y.: A comparison of prediction accuracy, complexity, and training time of thirty-three old and new classification algorithms. Machine Learning **40**(3), 203–228 (2000)
52. Liu, B., Hsu, W., Chen, S.: Using general impressions to analyze discovered classification rules. In: Proc. of the 3rd Int. Conf. on Knowledge Discovery and Data Mining (KDD-97), pp. 31–36. AAAI Press (1997)
53. Michalski, R.: On the quasi-minimal solution of the general covering problem. In: Proc. of the 5th Int. Symposium on Information Processing, pp. 125–128. Bled, Yugoslavia (1969)
54. Michie, D., Spiegelhalter, D.J., Taylor, C.C., Campbell, J. (eds.): Machine learning, neural and statistical classification. Ellis Horwood, Upper Saddle River, NJ, USA (1994)
55. Mingers, J.: An empirical comparision of pruning methods for decision tree induction. Machine Learning **4**, 227–243 (1989)
56. Murthy, S.: Automatic construction of decision trees from data. Data Mining and Knowledge Discovery **2**(4), 345–389 (1998)
57. Murthy, S.K., Salzberg, S.: Lookahead and pathology in decision tree induction. In: Proc. of the 14th Int. Joint Conf. on Artificial Intelligence (IJCAI-95), pp. 1025–1033 (1995)
58. Nuñez, H., Angulo, C., Catala, A.: Rule extraction from support vector machines. In: Proc. of the European Symposium on Artificial Neural Networks (ESANN-02), pp. 107–112 (2002)
59. Ohsaki, M., Kitaguchi, S., Okamoto, K., Yokoi, H., Yamaguchi, T.: Evaluation of rule interestingness measures with a clinical dataset on hepatitis. In: Proc. of the 8th European Conf. on Principles and Practice of Knowledge Discovery in Databases (PKDD-04), pp. 362–373. Springer-Verlag New York (2004)
60. Pazzani, M.J.: Knowledge discovery from data? IEEE Intelligent Systems **15**(2), 10–13 (2000)
61. Pfahringer, B., Bensusan, H., Giraud-Carrier, C.: Meta-learning by landmarking various learning algorithms. In: Proc.of the 17th Int. Conf. on Machine Learning, (ICML-00), pp. 743–750. Morgan Kaufmann, San Francisco, California (2000)
62. Provost, F., Kolluri, V.: A survey of methods for scaling up inductive algorithms. Data Mining Knowledge Discovery **3**(2), 131–169 (1999)

63. Prudencio, R., Ludemir, T.: Active selection of training examples for meta-learning. In: Proc. of the 7th Int. Conf. on Hybrid Intelligent Systems, pp. 126–131. IEEE Press (2007)
64. Quinlan, J.R.: Simplifying decision trees. International Journal of Man-Machine Studies **27**, 221–234 (1987)
65. Quinlan, J.R.: Learning logical definitions from relations. Machine Learning **5**, 239–266 (1990)
66. Quinlan, J.R.: C4.5: programs for machine learning. Morgan Kaufmann (1993)
67. Raedt, L.D.: Logical and Relational Learning. Springer (2008)
68. Romao, W., Freitas, A.A., Gimenes, I.M.S.: Discovering interesting knowledge from a science and technology database with a genetic algorithm. Applied Soft Computing **4**, 121–137 (2004)
69. Schaffer, C.: Overfitting avoidance as bias. Machine Learning **10**(2), 153–178 (1993)
70. Smyth, P., Goodman, R.M.: An information theoretic approach to rule induction from databases. IEEE Transactions on Knowledge and Data Engineering **4**(4), 301–316 (1992)
71. Szafron, D., Lu, P., Greiner, R., Wishart, D., Poulin, B., Eisner, R., Lu, Z., Poulin, B., Anvik, J., Macdonnel, C.: Proteome analyst – transparent high-throughput protein annotation: function, localization and custom predictors. Nuclei Acids Research **32**, W365–W371 (2004)
72. Theron, H., Cloete, I.: BEXA: A covering algorithm for learning propositional concept descriptions. Machine Learning **24**(1), 5–40 (1996)
73. Tsumoto, S.: Clinical knowledge discovery in hospital information systems: two case studies. In: Proc. of the 4th European Conf. on Principles of Data Mining and Knowledge Discovery (PKDD-00), pp. 652–656. Springer-Verlag, London, UK (2000)
74. Vilalta, R., Drissi, Y.: A perspective view and survey of meta-learning. Artificial Intelligence Review **18**(2), 77–95 (2002)
75. Vilalta, R., Giraud-Carrier, C., Brazdil, P., Soares, C.: Using meta-learning to support data mining. International Journal of Computer Science and Applications **1**(1), 31–45 (2004)
76. Weiss, S., Kulikowski, C.: Computer Systems that Learn. Morgan Kaufmann (1991)
77. Weiss, S.M., Indurkhya, N.: Optimized rule induction. IEEE Expert: Intelligent Systems and Their Applications **8**(6), 61–69 (1993)
78. Witten, I.H., Frank, E.: Data Mining: Practical Machine Learning Tools and Techniques with Java Implementations, 2nd edn. Morgan Kaufmann (2005)
79. Zyl, J.V., Cloete, I.: Fuzzconri – a fuzzy conjunctive rule inducer. In: J. Fürnkranz (ed.) Proc. of the ECML/PKDD-2004 Workshop on Advances in Inductive Learning, pp. 548–559. Pisa (2004)

Chapter 3
Evolutionary Algorithms

3.1 Introduction

Evolutionary Algorithms (EAs) are stochastic search methods inspired by the Darwinian concepts of evolution and survival of the fittest. They became popular methods for solving many kinds of problems, such as function optimization and several data mining tasks, including classification. EAs perform a robust global search in the space of candidate solutions, and are also well known for their associated implicit parallelism and noise tolerance [4, 29]. In addition, their global search performs a wider exploration of the search space than the greedy, local search performed by many other search methods, making EAs less likely to get trapped in local optima (suboptimal solutions) in the search space than local search methods. On the other hand, it should be noted that EAs tend to consume more computation time than local search methods.

In essence, an EA evolves a population of individuals, where each individual is a candidate solution for the target problem. At each generation, the individuals are evaluated according to a fitness function. The best individuals are more likely to be selected to reproduce and to undergo crossover and mutation procedures, to produce new offspring (new candidate solutions) that inherit some features from their parents. The evolutionary process is iteratively performed until a stopping criterion is satisfied, such as a maximum number of generations (iterations) being reached or an optimal solution being found.

According to the representation of their individuals and the evolutionary operators they use, EAs are often classified as genetic algorithms, genetic programming, evolutionary strategies, and evolutionary programming [4]. In this work we are particularly interested in genetic programming (GP), which will be discussed in detail in Sections 3.5 and 3.6. Before we discuss GP, however, Section 3.2 presents an overview of evolutionary algorithm concepts and principles. Section 3.3 reviews the main concepts of multiobjective optimization, and Section 3.4 explains the main differences between genetic algorithms and genetic programming. Finally, a summary of this chapter is presented in Section 3.7.

G.L. Pappa, A.A. Freitas, *Automating the Design of Data Mining Algorithms*,
Natural Computing Series, DOI 10.1007/978-3-642-02541-9_3,

3.2 An Overview of Evolutionary Algorithms

This section introduces the four main elements found in any evolutionary algorithm: individual representation, fitness function, selection methods, and genetic operators. It is important to emphasize that the concepts of individual representation and genetic operators are dependent on the type of evolutionary algorithm being used. Genetic algorithms' individuals usually are represented by fixed-length linear strings, while genetic programming's individuals usually are represented by variable-size trees, with corresponding differences in the genetic operators. In this section, the discussions on individual representations presented in Section 3.2.1 and genetic operators presented in Section 3.2.4 are in the context of simple fixed-length, linear-string individual representations, typically used in genetic algorithms. GP's tree-based individual representations and corresponding genetic operators will be discussed in Section 3.5.

In contrast, the concepts of fitness function and selection methods, presented in Sections 3.2.2 and 3.2.3, respectively, are more general, and do not depend on the type of evolutionary algorithm being implemented.

3.2.1 Individual Representation

Recall that in evolutionary algorithms, in general, an individual is a candidate solution to the target problem. In Genetic Algorithms (GAs) an individual (often called a "chromosome" by analogy with its biological counterpart) is often represented by a fixed-length string of genes, where each gene is a small part of a candidate solution.

In the simplest case an individual is represented by a string of bits, so that each gene can take on a value either 0 or 1. This representation is natural in problems involving binary decisions only. For instance, consider the well-known attribute selection (or feature selection) task of data mining [49], where the goal is to select a subset of relevant attributes out of all available attributes in the data being mined. In this type of problem a candidate solution can be naturally described by a binary string with n genes, where n is the number of attributes, and each gene takes on the value 0 or 1 to indicate whether or not its corresponding attribute is selected. Examples of GAs for attribute selection can be found, for instance, in [22, 44, 50, 61], and a brief overview of a number of such GAs can be found in [26], where an alternative individual representation is mentioned for attribute selection and the diversity of criteria used in the fitness functions of GAs for attribute selection is shown.

However, the binary representation has some drawbacks when it is used to represent numerical variables (taking real or integer values). One such drawback is the "Hamming cliff" effect [68]. This refers to the fact that two decoded integer values that are very similar to each other can have very different representations in binary strings. For instance, the integer numbers 15 and 16 are represented in conventional binary notation as 01111 and 10000, respectively. This reduces the "locality" of the representation, in the sense that a small local change in a single bit of the individual's genotype might have a large effect on its decoded candidate solution (its "phenotype"), which might reduce the effectiveness of genetic operators that assume the locality of the individuals' representation. The Hamming cliff effect can be avoided by using a Gray encoding, a clever approach for encoding integer numbers into a binary representation in a way that the average contribution of each bit to the represented integer is the same as that for each bit in the individuals' representation; see [68] for details.

It should be noted that the previously mentioned drawback of Hamming cliffs can be avoided by the direct encoding of real or integer values into the individuals, so that each gene is a real or integer value according to the data type of the variables in the target problem. More generally, in many real-world applications the need for individual representations that are closer to the problem being solved, involving problem-specific data structures, has been recognized for a long time [53].

For a more comprehensive discussion of individual representations for evolutionary algorithms, the reader is referred to a specialized research-oriented book by Rothlauf [68].

3.2.2 Fitness Function

The fitness function is a crucial component of an evolutionary algorithm (EA). The fitness function is used to evaluate how well an individual solves the target problem, and it is responsible for determining which individuals will reproduce and have parts of their genetic material (i.e., parts of their candidate solution) passed onto the next generation. The better the fitness of an individual, the higher the probability of that individual being selected for reproduction, crossover, and mutation operations.

Ideally, the fitness function should measure the quality of an individual (candidate solution) as precisely as possible, but the design of a fitness function should also take into account restrictions about available processing power, background knowledge about the target problem, and the user's requirements. In any case, it is important that the values of the fitness function be graded in a fine-grained manner; otherwise, the EA will have little information about individual quality to guide its search. For instance, in one extreme, a very bad fitness function would take on only the value 1 or 0, corresponding to "good" or "bad" individuals, and the EA would be unable to effectively select from among many different "good" individuals.

It is well known that, in most of the real-world problems addressed by an EA or other search method, the quality of a candidate solution should be evaluated by multiple criteria rather than a single criterion, which should lead to a multiobjective optimization problem. A detailed discussion of the approaches that can be followed to tackle this problem can be found in Section 3.3.

3.2.3 Individual Selection

Recall that the basic idea of selection in evolutionary algorithms is that the better the fitness (quality measure) of an individual (candidate solution), the higher must be its probability of being selected, and so the greater the amount of its genetic material that will be passed on to future generations. In this section we present a very brief overview of two popular selection methods, namely proportionate selection and tournament selection.

Proportionate selection is usually implemented by a computational simulation of a biased roulette wheel [29], and for this reason it is often called roulette wheel selection. Think of each individual in the population as a slot of a roulette wheel. This roulette wheel is biased in the sense that the size of a slot is proportional to the fitness of the corresponding individual. In order to select an individual, one simulates the spinning of the roulette wheel, so that the probability of an individual being selected is proportional to its fitness. The probability of an individual being selected is given by the ratio of its fitness value (analogous to its roulette wheel slot size) to the sum of fitness values of all population individuals (analogous to the entire size of the roulette wheel).

Tournament selection consists of randomly choosing k individuals from the population – where k is a user-defined parameter called the tournament size – and letting them "play a tournament" [10]. In general, only one winner from among the k participants is selected, where the winner is the individual with the best fitness out of all the k tournament players.

In order to choose a selection method and to instantiate its parameters, it is useful to consider its selective pressure. The selective pressure of a selection method can be defined as "... the speed at which the best solution in the initial population would occupy the complete population by repeated application of the selection operator alone" [18, p. 170]. If the selective pressure is too strong the population would probably converge prematurely to a local (rather than the global) optimum in the search space. If the selective pressure is too weak the search would be strongly random, since the best (fittest) individuals would be selected with probabilities similar to the probabilities of less fit individuals.

Selective pressure can be controlled in the tournament selection method by adjusting the value of the parameter k, the tournament size. The larger the value of k, the stronger the selective pressure.

3.2.4 Genetic Operators

The two most used types of genetic operators are crossover and mutation. The crossover (or recombination) operator essentially swaps genetic material between two "parents," creating two new "child" individuals. In the context of the fixed-length, linear-string individuals typically used in GAs, crossover is usually a simple operator. Figure 3.1 illustrates a simple type of crossover called one-point crossover between two parents, each of them with five genes. Figure 3.1(a) shows the parents before crossover. A crossover point is randomly chosen, represented in the figure by a vertical bar (|) between the third and fourth genes. Then the genes to the right of the crossover point are swapped between the two parents, yielding the new child individuals shown in Fig. 3.1(b).

$X_1 \quad X_2 \quad X_3 \mid X_4 \quad X_5$ $\qquad$ $X_1 \quad X_2 \quad X_3 \mid Y_4 \quad Y_5$

$Y_1 \quad Y_2 \quad Y_3 \mid Y_4 \quad Y_5$ $\qquad$ $Y_1 \quad Y_2 \quad Y_3 \mid X_4 \quad X_5$

(a) Before crossover (b) After crossover

Fig. 3.1 Simple example of one-point crossover in genetic algorithms

Another form of crossover that can be applied to fixed-length, linear-string individuals is uniform crossover [24, 72]. In this kind of crossover, for each gene position, the genes from the two parents are swapped with a fixed, position-independent probability p. The effect of this kind of crossover is illustrated in Fig. 3.2, where the three swapped genes are denoted by a box. Note that the swapped genes do not need to be adjacent to each other, unlike what happens in one-point crossover. Note also that, in general, the closer the gene-swapping probability p is to 0.5, the larger the number of genes swapped between the two parents, and so the greater the exploratory power of the crossover, i.e., the more global the search performed by this operator. Conversely, the closer p is to 0 or 1, the smaller the number of genes swapped between the two parents, and so the more local the search performed by this operator.

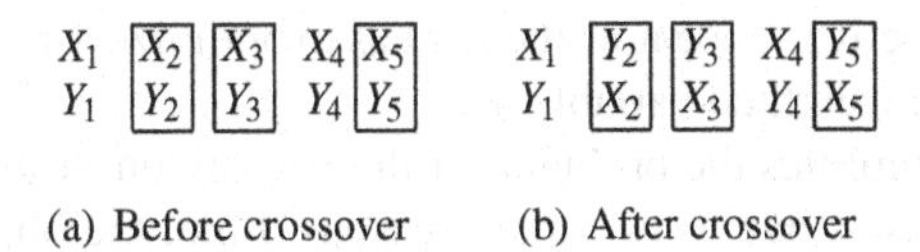

Fig. 3.2 Simple example of uniform crossover in genetic algorithms

Any crossover operator, as well as any other genetic operator, has a search bias, whose effectiveness depends on the application domain [27]. In particular, one-point crossover has a strong positional bias [23], since the probability of two adjacent gene values being swapped together to produce offspring is much higher than the probability of two very distant genes being swapped together. Hence, the positions of the genes within the individuals will affect which offspring are produced by the operator. Note that uniform crossover has no such bias, since it swaps gene values independently of their position in the individuals. On the other hand, uniform crossover has a strong distributional bias, in the sense that the number of genes that is expected to be swapped between the two parents is distributed around a certain value (given by the gene-swapping probability p multiplied by the number of genes in the individuals). Note that one-point crossover has no such bias, since all the possible numbers of genes to be swapped are equally likely, given that the crossover point is randomly chosen.

In GAs, mutation is an operator that acts on a single individual at a time. Unlike crossover, which recombines genetic material between two (or more) parents, mutation replaces the value of a gene with a randomly generated value. In the simplest individual representation, where an individual consists of a fixed-length, binary string, mutation simply inverts the value of a gene (a bit), i.e., it replaces a "0" with a "1" or vice versa. More complex types of mutation operators are needed to cope with more complex individual representations [53]. In GAs, mutation is usually applied with a small probability, typically much smaller than the crossover probability.

3.3 Multiobjective Optimization

The majority of problems solved by Evolutionary Algorithms (EAs) involves optimizing more than one aspect of the problem simultaneously. In this multiobjective optimization scenario, there are essentially three approaches that can be followed when designing a fitness function.

The first approach consists of assigning a user-defined weight to each objective (quality criterion to be optimized), indicating the relative importance of that objective, and combining all the weighted objectives into a single formula. This effectively transforms the original multiobjective problem into a single-objective one from the EA's perspective, since now the EA has to optimize a single fitness function (containing multiple weighted variables).

This approach simplifies the problem for the EA, but unfortunately it is not very effective in many cases, due to two main reasons. First, the objectives being optimized are often conflicting with each other, and by specifying a fixed weight for each objective a priori (before any search is done), the EA would be searching for an optimal solution according to just one possible trade-off between the conflicting objectives. Many other possible trade-offs between the objectives are ignored, so this approach heavily relies on the user to specify a good set of weights for the objectives. Secondly, the objectives often represent different and non-commensurate aspects of a candidate solution's quality, so mixing them into a single formula is not semantically meaningful.

The other two approaches for coping with multiobjective optimization avoid the need to transform a multiobjective fitness function into a single-objective function. These two approaches are based on Pareto optimality (or Pareto dominance) and on lexicographic optimization.

3.3.1 The Pareto Optimality Concept

According to the Pareto's multiobjective optimization concept, when many objectives are simultaneously optimized, there is no single optimal solution. Rather, there is a set of optimal solutions, each one considering a certain trade-off among the objectives [16]. A system developed to solve this kind of problem returns a set of optimal solutions, and it is left to the user to choose the one that best solves his or her specific problem. This means that the user has the opportunity of choosing the solution that represents the best trade-off among the conflicting objectives a posteriori, i.e., after examining several high-quality solutions. Intuitively, this is better than forcing the user to define a single trade-off between conflicting objectives by specifying a numerical weight to each objective a priori, i.e., before the search is performed. The latter scenario is what happens when the multiobjective problem is transformed into a single-objective one by assigning weights to each objective and combining all objectives into a single weighted formula.

The Pareto's multiobjective optimization concept is based on the Pareto dominance concept, which is used to guide the search for a set of optimal, non-dominated solutions. According to the Pareto dominance concept, a solution S_1 dominates a solution S_2 if and only if [19]:

- Solution S_1 is not worse than solution S_2 in any of the objectives;
- Solution S_1 is strictly better than solution S_2 in at least one of the objectives.

Figure 3.3 shows an example of possible solutions found by an EA when solving a rule induction problem where there are two objectives: to minimize both the classification error rate and the total number of rule conditions present in a rule set. The solutions that are not dominated by any other solutions are estimated to be Pareto-optimal solutions, and examples of such solutions are shown along the dotted line in Fig. 3.3. Note that Solution A has a small error rate but a large number of rule conditions. Solution B has a large error rate but a small number of rule conditions. According to the Pareto optimality concept, one cannot say that solution A is better than B, or vice versa, since neither of them dominates the other. (This point will be revisited in the next subsection.) On the other hand, solution C is clearly not a good solution, since it is dominated by solution B.

In an EA, the use of a multiobjective fitness function based on Pareto dominance allows the algorithm to return to the user the best approximation to the Pareto front found by the algorithm along its search. The user can then choose the best solution based on his or her desired trade-off between the objectives. In cases where the system does not have a direct user or the user is not able or willing to make a decision, an automated decision-making process can choose one of the non-dominated solutions over the other non-dominated ones, although this is far from a trivial problem. An example of such an automated approach is found in [61], where the use of a decision-making criterion based on the number of individuals in the population that are dominated by each solution in the estimated Pareto front is suggested. The more solutions (individuals) in the final population a given solution S in the estimated

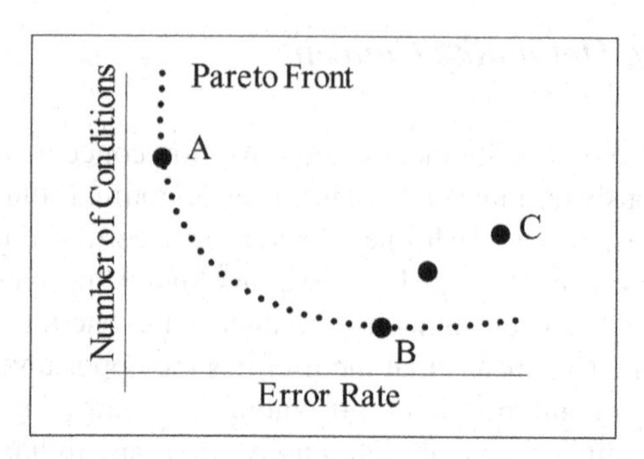

Fig. 3.3 Pareto front obtained when optimizing both the error rate and the total number of rule conditions produced by a rule induction algorithm

Pareto front dominates, the larger the chance of solution S being selected as the solution to be returned to the user. A review of other methods for decision making in multiobjective problems can be found in [16].

The use of multiobjective fitness based on Pareto optimality is becoming more and more common in EAs [15, 16, 19], as well as in data mining in general [28, 37]. An overview of a number of GP algorithms using Pareto optimality-based multiobjective fitness functions is provided in [65].

3.3.2 Lexicographic Multiobjective Optimization

The lexicographic approach for multiobjective optimization has been much less used in practice than the Pareto optimality-based approach. This is unfortunate, since the former does have some advantages over the latter in some applications, as will be discussed below.

In essence, in the lexicographic approach, the objectives are sorted in decreasing order of importance, according to the user's preferences, and the system tries to optimize the objectives in that specified order or priority [28, 39].

It is interesting to note that the lexicographic multiobjective approach can be regarded as an intermediate approach between the two "extreme approaches" of using a weighted formula and the Pareto optimality concept, as follows.

As discussed earlier, the weighted-formula approach requires the user to specify a numerical weight for each objective a priori, which puts a large burden on the user to specify the right weights and makes the system focus on a single trade-off between the different objectives. In contrast, in the Pareto optimality approach, the user does not specify any preference for any of the objectives, so the system has the very wild task of considering all possible trade-offs between the objectives.

In the lexicographic approach, the system uses information on the relative priorities of the objectives provided by the user. However, instead of having to specify ad hoc numerical weights for each objective, the user just has to rank the objectives in decreasing order of importance, i.e., the user provides only qualitative (rather than quantitative) information about the objective's priorities. In many applications this is intuitively considerably easier for users, and most users would agree on a given ordering of objectives, even though they would significantly differ in their opinions on the exact numerical weights. Note that, if the user has a clear idea about the order of priority of the objectives, the Pareto optimality approach would ignore this valuable information, and it would be too "conservative," since it would return to the user many non-dominated solutions whose objective values are unsatisfactory according to the user's objective priorities.

As an example of a scenario where the lexicographic approach seems to be more intuitively desirable than the Pareto optimality-based approach, consider the problem of discovering a set of classification rules where the two objectives to be minimized are the classification error rate and the total number of rule conditions in the rule set. Assume that, other things being equal, smaller rule sets are preferable to larger rule sets, which is justified by the well-known principle of Occam's Razor; see [20, 76]. The Pareto optimality-based approach would implicitly treat both objectives as equally important, without recognizing their relative priorities. However, although it is difficult to justify the use of a given precise numerical weight for each of those two objectives, most data mining researchers and practitioners would agree that minimizing the classification error rate is more important than minimizing the total number of conditions in a rule set. Consider now two candidate solutions A and B such that A has a significantly lower error rate but a significantly larger number of rule conditions than B, as illustrated in Fig. 3.3. In this case the system cannot prefer one of these two solutions over the other based on Pareto optimality, since neither of them dominates the other. However, the lexicographic approach would clearly identify solution A as better than solution B, in agreement with the user's preferences. An example of a work proposing an EA for data mining based on lexicographic multiobjective optimization can be found in [9].

3.4 Genetic Programming Versus Genetic Algorithms: A Critical Perspective

Genetic Programming (GP) [8, 42] is an area of evolutionary computation where the goal is to automatically evolve computer programs. It first became popular for evolving a population of trees representing Lisp S-expressions, and it has been applied to the evolution of candidate solutions for a wide range of problems [42, 43]. As research in the area of GP has developed, many other representations for GP individuals have been proposed, including GP individuals with GA- (Genetic Algorithm-) like genomes (fixed-length linear genomes), individuals represented by graphs, and based on grammars. Programs in a variety of languages other than Lisp have also been created (including Prolog, C, and Java).

In this book we will focus on the tree-based representation and one of its extensions, namely a grammar-based tree representation where the trees representing

Algorithm 3.1: Genetic algorithms and genetic programming

```
Create initial population of individuals
Compute fitness of each individual
repeat
    Select parent individuals based on fitness
    Apply genetic operators to selected parent individuals, creating child individuals
    Compute fitness of each child individual
    Update the current population
until stopping criterion
```

individuals are created by applying the production rules of a grammar. The conventional tree-based representation is used by the vast majority of GP algorithms in the literature, and standard GP using this representation is discussed in Section 3.5. Grammar-based GP is discussed in Section 3.6. Before we discuss GP in more detail, though, we must discuss the similarities and differences between GA and GP, because these are very important for understanding the core contribution of this book, namely our proposed GP system to automatically design a data mining algorithm, which will be described in detail in Chapter 5.

The first point to note is that, at a very high level of abstraction, GP and GA can be described by exactly the same pseudocode as that shown in Alg. 3.1. There are, however, significant differences in the types of individual representations, genetic operators, and the fundamental nature of the solution produced by each of these two types of evolutionary algorithms, which are not shown in such high-level pseudocode.

Let us consider first the issue of individual representations. As discussed earlier, most GAs use a fixed-length linear-string individual representation, while most GP algorithms use a variable-size tree-based representation. Unlike what happens in "conventional" GAs, in GP an individual's tree can grow in size and shape in a very dynamical way as a result of the evolution of the population. Another difference between GA and GP, associated with their individual representations, concerns the genetic operators they use. The use of tree-based representation in GP require somewhat more complex crossover and mutation operators than the use of linear representations in GAs, as will be discussed later.

Arguably, however, the above two differences are not so important as criteria for distinguishing between GA and GP. There are variants of GAs that use a variable-length individual representation (requiring more sophisticated generic operators), e.g., [36] and the variable string length GA classifier described in [5], which addresses the classification task of data mining; and there are variants of GP that use a fixed-length linear individual representation, e.g., [48, 83], which describe gene expression programming algorithms for the classification task. In addition, whether the candidate solutions being evolved have fixed length or variable length is mainly a syntactical issue that is relatively less important than semantical issues related to the nature and scope of a candidate solution. Let us elaborate on this latter point.

In principle, GAs and GP algorithms are supposed to have individuals that represent fundamentally different types of candidate solutions with respect to generality and expressiveness power. GAs normally use individual representations with a relatively low expressiveness power, where the individual genes typically correspond to data such as values of variables. In contrast, GP algorithms normally use individual representations with considerably greater expressiveness power, where an individual contains both data (typically terminal nodes in the individual's tree) and functions or operators (typically nonterminal nodes in the individual's tree). The ability to evolve candidate solutions expressed in terms of both data and functions (or operators) is a very important characteristic of GP, which allows it to discover solutions to more complex problems than other conventional types of evolutionary algorithms.

This ability is also a significant step towards the evolution of programs, the defining goal of GP. However, the fact that a candidate solution is expressed in terms of both data and functions (or operators) does not necessarily make it a "program" in the normal sense of the term in computer science, a point that seems to be missed or ignored by a large part of the GP literature. A computer program normally also involves programming constructs such as (potentially nested) conditional (*if-then-else*) and loop (*while, for or repeat-until*) statements. Although many GP algorithms use some *if-then-else* statements in their individual representation, sometimes such statements are used in a simplified form, allowing just one such statement per individual (the *if* statement is often put in the root node of all individuals to enforce a certain kind of structure in the candidate solutions), and not allowing nested *if-then-else* statements. More importantly, the proportion of GP algorithms in the GP literature using individual representations with loops is quite low, a fact disappointing for computer scientists in general, given the fundamental nature of loops in computer programming.

Concerning the generality of candidate solutions, a crucial point is that GAs aim at finding a solution for a specific instance of the target problem, while GP algorithms aim (or at least should aim) at finding a computer program (an algorithm, a generic "recipe") that represents a general solution of the target problem, which could be applied to any instance of that problem. This is arguably *a point very often missed or ignored in the GP literature*, so let us discuss it in more detail, referring to two very different types of problems as examples.

Let us consider first the classical and very well-known Traveling Salesman Problem, a type of combinatorial optimization problem. In essence, this problem can be defined as follows. We are given a graph with a set of nodes and a set of edges, where each node corresponds to a city and each edge connecting two nodes represents a direct path between the corresponding two cities. Each edge has a numerical weight indicating the distance between its two cities. A complete, fully connected graph is normally assumed, where there is an edge between every pair of nodes. The goal is to find the Hamiltonian tour – a path that starts in one city, passes exactly once through every other city, and returns to the start city – of shortest length from among all possible tours.

Many GAs have been proposed to solve the Traveling Salesman Problem and related problems where a candidate solution consists of essentially a permutation of labels [21, 64, 70, 74]. However, it should be noted that these GAs try to find an optimal solution to one particular instance of this problem, i.e., the GA receives as input a particular set of cities and weighted edges, and outputs a particular tour – hopefully a (near-)optimal one – for that particular instance of the problem.

In contrast, since GP is or should be about the evolution of generic programs, a GP algorithm should produce a generic algorithm for solving any instance of the Traveling Salesman Problem. In order words, the output of the GP algorithm should be a generic program that is capable of (a) receiving as input any set of cities and weighted edges corresponding to any instance of a Traveling Salesman Problem, and (b) producing as output a hopefully (near-)optimal tour for that input. The fundamental differences between the generality and expressiveness power of the type of candidate solutions produced by existing GAs and the candidate solutions that should be produced by GP algorithms for the Traveling Salesman Problem are clearly shown in Fig. 3.4, where the implementation of the "choose-next-city-to-visit" procedure would have to be evolved by the GP algorithm.

London Paris Madrid ... Brussels London

(a) Type of solution produced by Genetic Algorithms, a solution (list of cities) for an instance of the TSP

```
Initialize tour with randomly-chosen city
Initialize tour cost with 0
Mark all cities as unvisited
WHILE (there is an unvisited city)
  Perform procedure Choose-next-city-to-visit
  Mark chosen city as visited and add it to tour
  Update tour cost
END WHILE
Return tour
```

(b) Type of solution that should produced by GP, a generic program to solve any instance of the TSP

Fig. 3.4 Very different types of candidate solutions for the Traveling Salesman Problem (TSP)

Let us consider now a very different type of problem as another example of the fundamental differences between the types of candidate solutions produced by GAs and GP algorithms. One of the most common applications of GAs is function optimization, where a GA is given a well-defined function involving one or more numerical variables, and it has to find the values of the variables that lead to the optimal (maximal or minimal) value of the function. A simple example of a function to be optimized is $sin(x) - 0.1x + 2$, in which case a candidate solution would consist simply of a value of the variable x. Since the nature of a candidate solution involves only data (variable values), and not functions or operators, there is no need for GP algorithms to solve this function optimization problem.

On the other hand, GP algorithms are often used to solve a related but usually more difficult problem, namely the problem of function approximation. In this problem, the GP algorithm is given a set of pairs of variable values and their corresponding function values, i.e., pairs of the form $(x_i, f(x_i))$, for i=1 ... n, where n is the number of available data points. Note that the GP algorithm is not told the definition of the function $f(x)$. Actually, in this type of problem the goal of the GP algorithm is to find out which function best matches (or approximates) the input data points, which is essentially a function approximation problem, also called a regression problem. This type of problem is usually called symbolic regression in the GP literature, which indicates that the GP has to find a symbolic expression that best approximates the unknown target function. This type of symbolic regression problem will be discussed in somewhat more detail later. For now the main point is that in this type of problem a candidate solution consists of the mathematical specification of a function, which includes both data (variables and constants) and mathematical operators. Hence, this type of candidate solution representation has considerably more expressiveness power than the simpler candidate solution representation used by GAs to solve a function optimization problem, which includes only data.

Despite its use of both data and functions (or operators) in the representation of candidate solutions, the aforementioned conventional type of GP algorithm for symbolic regression should not be considered as the ideal or prototypical GP algorithm for generating very generic programs, as discused earlier, for two reasons. First, in general the individuals of such GP algorithms do not contain loops and other typical program constructs such as nested *if-then* statements; their individuals essentially represent a mathematical formula, though often a complex one. Secondly, and more importantly, the candidate solutions represented by the individuals of such algorithms are not very generic programs that can be applied to data from different application domains. To be fair, the evolved solutions have some type of generality because they can be applied to predict the value of a variable for any fitness case (data point used to "train" the GP algorithm) or even any other data point unseen during the run (or "training") of the GP algorithm, as long as the new data point comes from the application domain from which the fitness cases were drawn. However, the evolved solutions are just a regression model for that particular application domain or dataset. In contrast, a genuine GP algorithm should produce as output a truly generic program which could be applied to any regression dataset in any application domain, i.e., it could be a program with the same generality of a human-designed algorithm for regression (such as the well-known CART algorithm [11]).

3.5 Genetic Programming

In this section we discuss standard Genetic Programming (GP) concepts and methods. Grammar-based GP, a more sophisticated type of GP, is discussed in the next section and used extensively in the next chapters.

The design of a standard GP system involves specifying how to implement at least the following major components:

1. A set of functions and a set of terminals, used to create the GP individuals.
2. A representation for the individuals.
3. A population initialization method.
4. A fitness function, used to measure the quality of the individuals.
5. A selection method, which selects individuals based on their fitness values.
6. Crossover and mutation operators, which use the selected individuals to generate new offspring.

The terminal and function sets, together with the fitness function, are the most problem-dependent components of a GP algorithm, while the other components can be in general used to solve many different kinds of problems.

The function set and the terminal set define the basic elements with which a program (an individual) in GP is built. Terminals provide a value to the system, while functions process a value already in the system [8]. The terminal set is usually composed of constants, variables and/or zero-argument functions (e.g., a random-constant generator). In turn, the function set may include boolean and arithmetical functions, conditional and/or loop statements, and subroutines, among many other features. Since the terminal set can include zero-argument functions, Montana [54] suggested using the term "nonterminal set" instead of the term "function set." The former term seems more technically correct, but the latter term is used more in practice, so in this section we use the term function set to be consistent with the majority of the GP literature. In any case, in a tree-based representation, terminals are associated with leaf nodes, while functions and operators are associated with internal (non-leaf) nodes.

Once the set of functions and terminals are defined, they can be used to generate the first GP population. This population is formed by a set of individuals, where each individual represents a solution to the target problem. Hence, if we are developing a GP for discovering classification rules in a specific dataset, for instance, each GP individual will represent a single candidate rule or a set of candidate rules for that dataset.

There are two main types of individual representations used in the GP literature: the first one represents an individual as a tree, and the second as a linear structure (other types of representations, such as graphs, are also possible [8]). Figure 3.5 shows an example of a tree representation and a linear representation for a GP individual. Both representations encode the function $x^2 + 1$. In this example, * and + are functions, and x and 1 are terminals.

As observed in Fig. 3.5(a), in the tree representation the execution of the tree is usually made in postfix order (reading the leftmost node of the tree first), while a linear representation is simply a sequence of commands that are executed from left to right. Nonetheless, these conventions can be changed depending on the functions included in the function set.

After choosing the set of terminals and functions, and an individual representation, the next step in the implementation of a GP system is to create the first

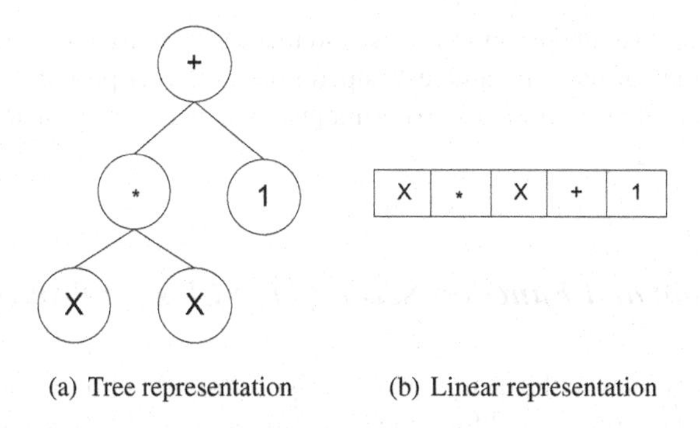

(a) Tree representation (b) Linear representation

Fig. 3.5 Standard GP representations

population. In the context of the very popular tree representation, there are three methods commonly used [42], named grow, full, and ramped-half-and-half – alternative methods are mentioned in [65].

The grow method creates individuals (with a predefined maximum depth) by selecting elements belonging to both the function and the terminal sets (except for the root node, which is selected from the function set). This selection scheme ends up producing trees with very diverse shapes and depths.

The full method, in turn, generates trees by selecting only elements from the function set, until the maximum depth of the tree is reached. At this point, it selects elements only from the terminal set to finalize the tree generation process. As a result of this initialization process, the trees produced have all the same depth.

In order to maintain diversity in the population of individuals generated, the ramped-half-and-half method combines the two just-described methods, as follows. Given the maximum depth M of a tree, the population is divided into equally distributed groups of individuals which will be initialized with a maximum depth varying from 2 to M. Within each of the created groups, half of the individuals is generated using the grow and half using the full initialization method.

It should be noted that any population initialization method has a bias that tends to be more suitable to some problems than to other problems. For instance, the full and the ramped-half-and-half methods tend to create bushy trees, where the leaves are on average at the same distance from the root of the tree, and this type of tree tends to be more suitable to problems involving some type of symmetry in the optimal solution, e.g., when all input variables are equally important, such as in multiplexer or parity problems [65].

At the end of the population initialization process, the individuals are evaluated according to a fitness function, and the ones representing the best candidate solutions for the problem are more likely to be selected to undergo reproduction, crossover, and mutation operations. Each of these operations is applied to the selected individuals according to a user-defined rate.

After the design of the main components of a GP algorithm is completed, we still have to choose values for standard parameters like population size, number of generations, crossover and mutation rates, and maximum individual size. In general, the optimal values of these parameters depend on the target problem, and they are usually chosen during preliminary experiments, as is usual in applications of most data mining methods.

3.5.1 Terminal and Function Sets and the Closure Property

One of the main issues in the design of a GP algorithm is to choose which elements will be used in the terminal and function sets. The optimal choice of these elements is of course very problem-dependent, but a generic, small set of logic and arithmetic functions seems to work reasonably well in many applications, as suggested by Banzhaf [8, p. 111]: "An approximate starting point for a function set might be the arithmetic and logic operators: PLUS, MINUS, TIMES, DIVIDE, OR, AND, XOR. The range of problems that can be solved with these functions is astonishing. Good solutions using only this function set have been obtained on several different classification problems, robotics control problems, and symbolic regression problems."

In any case, while choosing the elements that will compose the terminal and function sets, the GP designer has to keep in mind a trade-off between expressiveness power and computational efficiency. On the one hand, the terminal and function sets should have enough elements to be able to express a good candidate solution for the problem [42]. On the other hand, too many terminals and functions might create an unnecessarily large search space of candidate solutions, making it difficult to find a good solution in that space.

In addition, although the GP algorithm designer can freely choose the terminal and function sets used by the GP algorithm, the terminals and functions have to respect the closure property. This property states that every function in the function set has to be able to handle the values it receives as input (which can be a terminal or the output of another function). For instance, a division operator has to be modified to cope with division by zero. This is often implemented by making the operator return a given value, rather than an error, in the case of division by zero.

The constraints imposed by the closure property allowed the first GP systems to deal with only one data type. Nevertheless, new GP systems were developed in order to overcome this problem, as will be explained in Section 3.5.4.

Since the focus of this book is on data mining, let us discuss some issues in the design of the terminal and function sets particularly in the context of two types of data mining task, namely regression and classification. From a predictive data mining perspective, regression, like classification, can be considered a task involving the prediction of the value of a user-defined goal attribute for a new example (record, data instance) given the values of the other (predictor) attributes for that example. The main difference between classification and regression is as follows. In classification the goal attribute whose value has to be predicted is categorical (nominal), i.e., it can take on (unordered) nominal or discrete values, or classes, e.g. the two classes “good credit” and “bad credit.” In contrast, in regression the goal attribute is continuous (real-valued), so the prediction would be much more fine-grained, e.g., when trying to predict the exact amount of credit that a customer deserves, or to decide the maximum value of a loan that can be given to that customer.

Both in regression and in classification, an important issue affecting the design of the terminal and function sets is the set of data types of the predictor attributes. If all predictor attributes have the same data type, then it is easy to satisfy the aforementioned closure property by using, in the function set, only operators and functions that receive input arguments of that data type and output values of the same data type. For instance, if all predictor attributes are continuous (real-valued), one can simply include in the function set several kinds of mathematical functions appropriate to the application domain, and include in the terminal set the predictor attributes and a random-constant generator [42, Chap. 10]. When an example whose goal attribute value has to be predicted is given to the tree representing an individual, the predictor attributes at the leaves of the tree are instantiated with the corresponding predictor attribute values in that example. Once the system applies the functions in the internal nodes of the tree to the values of the attributes and randomly generated constants in the leaf nodes of the tree, it computes a numerical value that is output at the root node of the tree. If the target problem is regression, this output value is the goal-attribute value predicted by the GP algorithm for that example.

This same procedure can be used, with a relatively simple adaptation, when the target problem is classification, rather than regression. The basic idea of such an adaptation is to assign numerical values to the classes to be predicted and then predict the class whose value is closest to the numerical value output at the root of the tree. This procedure is most often used to discriminate between two classes, so that if the value output at the root of the tree is greater than a given threshold, the GP algorithm predicts a given class; otherwise, it predicts the other class.

If the dataset has a number C of classes such that $C > 2$, then the original problem is often transformed into C two-class problems [41], as follows. The GP is run C times. The ith run, $i = 1,...,C$, evolves an expression (individual) – a classification function – for discriminating between class c_i and all other classes. Then all C evolved expressions are used for predicting the class of new examples. Note that this requires some form of conflict resolution among the C evolved expressions when two or more expressions have their conditions satisfied by a given example. It is also possible to solve the C–class problem in a single run of the GP algorithm, rather than by running the GP algorithm C times. One approach to achieve this consists of using an individual representation based on multiple trees, so that, for a C–class problem, each individual consists of C trees, where each tree represents a classifier for its corresponding class [55].

Table 3.1 Example dataset for regression

x	−10	−8	−6	−4	−2	0	2	4	6	8	10
f(x)	153	91	45	15	1	3	21	55	105	171	253

3.5.2 Fitness Function: An Example Involving Regression

A discussion of the fitness function for GP in general was presented in Section 3.2.2. In this section we focus on a concrete example of a fitness function for a GP algorithm designed for solving a symbolic regression problem, which is a kind of data mining problem.

Recall that in the symbolic regression type of problem the GP algorithm is given a set of pairs of variable values and their corresponding function values, i.e., pairs of the form $(x_i, f(x_i))$, for i=1,...n, where n is the number of available data points. Each such pair of values is often called a "fitness case" in GP terminology. An example set of fitness cases for symbolic regression is shown in Table 3.1. Of course the fitness cases might have several variables (attributes), rather than just one variable, but in this example we will use only a univariate dataset for the sake of simplicity. Note that the GP algorithm is not told the definition of the function $f(x)$, since in this type of problem the goal of the GP algorithm is to find out which function best fits (or approximates) the fitness cases, as mentioned at the end of Section 3.4.

In order to compute the fitness of an individual for the regression problem given by Table 3.1, one needs to define a measure of the degree of fit between the symbolic expression represented by an individual and the fitness cases. A simple measure involves computing, for each fitness case i (i=1,...n), the squared error between the value of $f(x_i)$ computed by the individual's symbolic expression and the correct value of $f(x_i)$ given in Table 3.1. Once these n values are computed they are added up, giving the total sum of squared errors. To consider a concrete example, consider an individual representing the candidate solution $x^2 + 1$, as illustrated in Fig. 3.5, where both parts (a) and (b) of that figure represent the same candidate solution. The fitness of that individual would be given by the following summation:

$$(-10^2 + 1 - 153)^2 + (-8^2 + 1 - 91)^2 + \ldots + (10^2 + 1 - 253)^2$$

The above type of fitness function is often used in GP algorithms for the symbolic regression problem. However, it is important to note that this type of fitness function is measuring only the degree of fit between a candidate solution and the fitness cases used as input to the GP algorithm. There is nothing in this type of fitness function to encourage the discovery of a symbolic expression that has a good "generalization ability," i.e., an expression that would be able to predict the correct value of $f(x_i)$ not just for the fitness cases used during the run of the GP algorithm, but also for any other case to be observed in the future. This lack of consideration for generalization ability is also usually present in the GP literature for symbolic regression when evaluating the results of the GP algorithm, i.e., the final measure of performance involves the fitness of the best solution returned by the GP algorithm at the end of the GP run.

In the terminology of supervised data mining tasks, in such a scenario the evaluation of the performance of the GP algorithm is just using a "training set." Since no "test set" is used to measure the performance of the best solution returned by the GP algorithm, there is no predictive accuracy or generalization ability being measured. This considerably simplifies the problem from the point of view of designing a GP algorithm, but it also has the disadvantage of solving a less interesting (or at least easier) type of problem, since there is no need to worry about issues like overfitting, a difficult problem in predictive data mining (see Section 2.2.2).

Designing a GP algorithm for symbolic regression where the goal is to maximize the predictive accuracy in a test set containing cases unseen by the GP algorithm during its evolution is a considerably more challenging task; it involves incorporating in the GP algorithm mechanisms to cope with overfitting and related issues. Perhaps GP research would benefit from having more works consider such a challenging predictive regression scenario, instead of their focusing so much on the simpler regression scenario without prediction.

3.5.3 Selection and Genetic Operators

Following the evaluation of the GP individuals, a subset of them is selected to undergo reproduction, crossover and mutation operations. Selection methods are independent of the individual representation used by the GP algorithm, since these methods require, as input, only the individuals' fitness values, and not the individuals' contents. Hence, in principle a GP algorithm can use any selection method for evolutionary algorithms in general, such as the methods discussed in Section 3.2.3.

The individuals selected during the GP selection phase go through crossover, mutation, or reproduction operations according to user-defined rates. The reproduction operator simply copies the selected individual to the next generation without any alteration.

Crossover recombines the genetic material of two parent individuals in order to produce two new children. If the individuals are represented by trees, randomly selected subtrees are swapped between the two parents. In the case of linear genomes, randomly selected linear segments of code are swapped. A simple, abstract example of crossover in the case of the tree-based representation is shown in Fig. 3.6.

Note that, in the case of the tree representation, the internal nodes represent functions that usually have at least two input arguments, i.e., at least two child nodes, and so a tree representing a GP individual typically has many more leaf nodes than internal (non-leaf) nodes. As a result, if the crossover point (the root of a subtree being crossed over) is chosen at random with a uniform probability distribution, it is more likely that the crossover point will be a leaf node than an internal node of a parent. This is often undesirable, since in this case the crossover operator would degenerate into an operator that would tend to swap just leaves rather than subtrees,

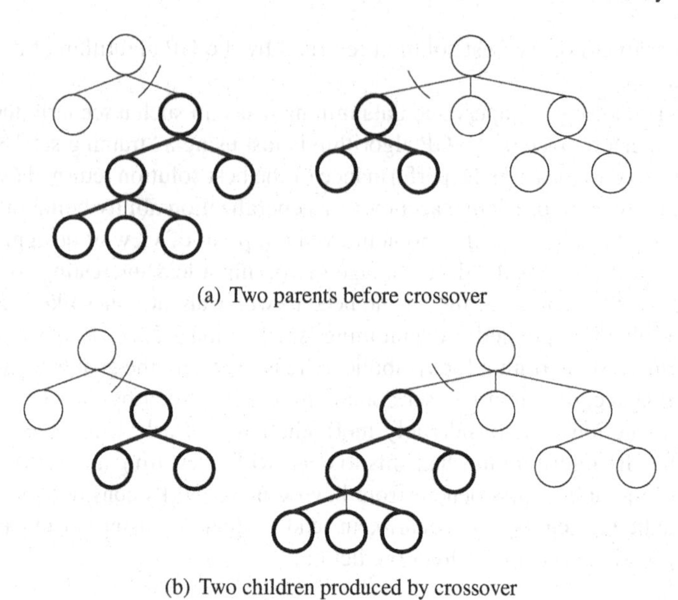

(a) Two parents before crossover

(b) Two children produced by crossover

Fig. 3.6 Example of tree crossover in genetic programming

significantly reducing the exploratory power of crossover. To avoid this, a common approach is to choose as a crossover point an internal node 90% of the time and a leaf node 10% of the time, as suggested by Koza [42].

Unlike crossover, mutation acts on a single parent individual of the population. One type of mutation consists of randomly selecting a subtree of the tree-based genome or a gene in a linear genome and replacing it by a new randomly generated subtree or gene. A simple example of mutation in the case of the tree-based representation is shown in Fig. 3.7.

Both crossover and mutation operations can be implemented in many ways [8]. Regardless of how they are implemented, they always have to respect the closure property, guaranteeing that all the newly generated individuals are valid.

In the GP community, a wide discussion about the effects of the crossover operator has taken place in the literature [2, 8, 51, 65, 82]. While in Genetic Algorithms (GAs) crossover has been considered a suitable method to preserve building blocks (good parts of candidate solutions), in GP its effectiveness has not been proved yet. Many researchers defend the idea that the crossover operator in GP systems is nothing more than a macromutation operator. According to empirical studies, crossover very often reduces the fitness of the offspring relative to their parents in the majority of GP algorithms [8].

One of the reasons for the bad results obtained with crossover operators in GP is related to the fact that they are context-insensitive. While in GAs crossover respects homology – because each gene has a specific function – in GP crossover does not take into account the context when choosing crossover points. Consider a

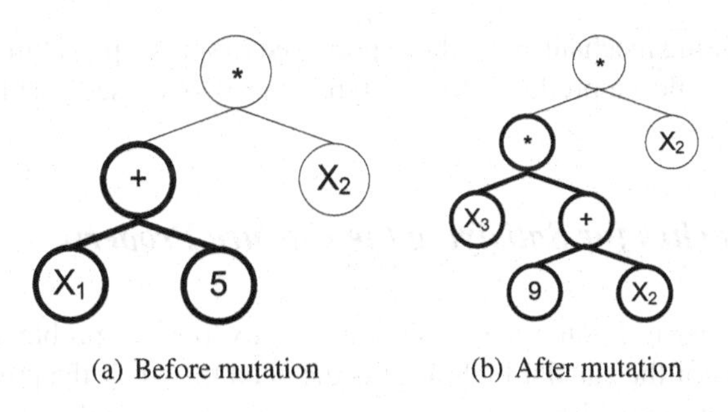

Fig. 3.7 Example of tree mutation in genetic programming

tree-represented GP individual, for example. A crossover point is randomly selected for each of the two individuals, and their genetic material swapped. However, a subtree that is good in the context in which it appears in individual 1 can be bad in the context in which it will appear in individual 2. This fact is not taken into account when standard GP crossover is performed. This problem led to the design of several homologous crossover operators for GP, and a good overview of such operators can be found in [65].

In any case, conventional crossover operators can be potentially improved by adding to them some kind of local search [82]. The basic idea is that, when two parents are crossed over, different crossover points are tried so that many (considerably more than two) potential children are produced. Each of these potential children has its fitness evaluated, and then only a selected subset of the fittest children (typically just the fittest or the two fittest) are selected to be passed to the next generation; the other potential children are simply discarded. For instance, in the context-aware crossover proposed in [51], first a crossover point is chosen at random in the first parent, as usual, but then the crossover procedure tries every possible crossover point in the second parent, i.e., for each possible crossover point in the second parent, the procedure swaps the corresponding subtrees between the two parents and evaluates their fitness. At the end of this process just the best child (out of all crossover attempts) is selected to be passed to the next generation, and the other potential children produced by the crossover procedure are discarded.

It should be noted that, by comparison with conventional crossover without local search, this type of local search procedure tends to improve the fitness of the offspring, but it also has the disadvantages of being in general more computationally expensive and more prone to premature convergence to a local (rather than the global) optimum in the search space, since this type of method effectively introduces a stronger selective pressure into the GP algorithm. After all, when local search is used in a crossover operator, the GP has two sources of selective pressure, the selection of the parents and the selection of the offspring in crossover.

Hence, when local search is used in crossover, it might be a good idea to lower the selective pressure in the method used to select the parents [82]. Concerning computational efficiency, in the particular case of the aforementioned context-aware crossover, a caching mechanism has been proposed to speed up the fitness evaluation of the potential children produced by that type of crossover (see [51] for details).

3.5.4 Approaches for Satisfying the Closure Property

Despite its success in finding good solutions for a variety of problems, one of the main drawbacks of the standard GP form is the need to satisfy the previously mentioned *closure* property. Recall that the closure property requires that all the GP terminals and GP functions produce a value that can be used as a valid input by another GP function. Because of this constraint, many conventional GP algorithms can work with only one kind of data type.

When the symbols in the terminal and function sets have different data types, a GP algorithm has to use some nonconventional approach to satisfy the closure property. There are at least three basic approaches for satisfying closure, as follows.

1. Converting the different data types of all terminals and all functions' inputs and outputs into a single standard data type, and then using a conventional GP algorithm coping with that data type only.
2. Using Strongly Typed Genetic Programming (STGP) [54].
3. Using Grammar-based Genetic Programming (GGP) [78].

The first approach, converting all data types into a single data type, is conceptually simple, while the other two approaches are more sophisticated. GGP will be discussed in detail in Section 3.6. Let us briefly discuss here the basic ideas of the STGP approach.

The STGP created by Montana [54] associates a data type with each terminal and each function's inputs and outputs, and the population initialization process and genetic operators are restricted – a child node has to return the data type expected by its parent node. This mechanism overcomes the problem of dealing with only one data type, but at the same time, as pointed out in [40], makes some zones in the search space inaccessible due to crossover restrictions. Nonetheless, one of the most interesting observations made in Montana's work is the fact that only 20 out of 50,000 individuals generated in the standard GP (i.e., not STGP) initial population of a multidimensional least squares problem were type-consistent. This observation emphasizes the importance of a strongly typed system in some applications.

3.5.5 Bloat

Bloat is essentially the uncontrolled large growth of individuals along the evolution without a corresponding improvement in fitness. Bloat is usually associated with an increase in the number of introns in individuals. An intron in the context of biology is essentially a DNA sequence, found in a gene, which is not used to create proteins.

In the context of GP algorithms, an intron is a sequence of (redundant) program code that does not directly affect the survivability of the individual [8], such as $x = x+0$ or $y = y*1$. As the GP search progresses, introns start growing exponentially (bloating), sometimes making the evolutionary process finish prematurely due to too much use of memory and/or processing time.

There are several hypotheses concerning the reasons for bloat, but so far there is no single hypothesis or unifying theory that is accepted by the entire GP community [65]. Here we discuss only two popular hypotheses, namely the "fitness causes bloat" hypothesis and the "neutral code is protective" hypothesis [7]. A more comprehensive discussion of these and other hypotheses can be found in [71].

The "fitness causes bloat" hypothesis can be summarized as follows [45, 46, 47]. When using variable-length individual representations, as is typically the case in GP, there are many ways of representing a good candidate solution. In general, for any given good candidate solution, there will be many more long representations than short ones. When fitness is based only on the quality of candidate solutions, and not on their length, all the representations of a given good candidate solution that have the same fitness value are equally likely. Hence, since there are more long representations than short ones for a given good candidate solution, the effect of fitness-based selection is to introduce a bias favoring longer representations. In addition, as the search progresses, more and more high-quality solutions are found, and so it becomes harder and harder for genetic operators to create offspring fitter than their parents. Hence, there is a selective pressure to produce offspring that at least have the same fitness as their parents, and a relatively easy way for the GP algorithm to do that is to keep sampling longer representations (with the same fitness) of the same basic high-quality candidate solution by adding redundant code to it, which is easier than creating a better candidate solution. The overall effect is that the size of the individuals in the populations grows fast, reaching a point where individuals have a large proportion of redundant code and the search tends to stagnation – unless some bloat-avoidance mechanism is used.

The "neutral code is protective" hypothesis [7] essentially states that bloat offers protection against the destructive effects of genetic operators such as standard tree crossover [2]. Recall that conventional crossover operators are usually "destructive" in the sense that they usually produce offspring with fitness lower than the fitness of their parents [8]. The larger the amount of redundant code of an individual, the more likely crossover will remove only the redundant code from that individual, and preserve its effective code. Although this is a popular hypothesis, according to [71] it has been developed mainly in the context of GP with linear representation and it may not be completely applicable to GP with tree representation.

In any case, it should be noted that, even if bloat offers some protection against the destructive effect of genetic operators such as conventional tree crossover, it does not necessarily follow that bloat is a good thing. After all, bloat has at least the disadvantage that computational resources such as memory and processing time are wasted by storing and manipulating large chunks of redundant code. More importantly, bloat leads to stagnation of the search for new good candidate solutions. Although a destructive genetic operator is not good, having a genetic operator that is not destructive but just produces larger representations of an already known good candidate solution is not particularly good either, since that genetic operator would hardly discover any new good solution. Hence, there is a clear motivation to try to avoid or at least to mitigate bloat in practice.

There are at least three basic approaches for avoiding or mitigating bloat [47, 65]. The first approach, the simplest one, consists of specifying a maximum size or depth limit for the individuals. Such a maximum size is usually specified in an ad hoc, static fashion. This approach can be made more flexible, though. For instance, one can use a "soft upper bound" for the size of an individual [73], where individuals larger than that upper bound are not removed from the population, but rather are more likely to undergo a mutation that will reduce their size; or one can use a dynamic individual depth or size limit that can be raised or lowered, depending on the best candidate solution found so far by the GP algorithm [71].

The second approach consists of using a fitness function that measures not only the quality of an individual but also its size, favoring of course smaller individuals. This leads to multiobjective fitness functions, as discussed in Section 3.2.2. An overview of several GP algorithms using a multiobjective fitness function to avoid or reduce bloat (where one of the objectives to be minimized is the size of an individual) can be found in [65].

The third approach involves designing a new genetic operator specifically tailored to avoid or mitigate bloating. To consider a simple example, a mutation operator that always reduces the size of an individual by replacing a subtree that has more than one node with a single leaf node would have a bias towards smaller trees, being therefore a very simple way of fighting bloat.

3.6 Grammar-Based Genetic Programming

This section introduces Grammar-based GP (GGP), a variation of the standard GP described in the previous section. As the name suggests, the main difference between the standard GP approach and a grammar-based one is the presence of a grammar. In GGP systems, the set of terminals and functions is replaced by a grammar. The grammar guarantees that all the individuals are syntactically correct. Note that in GGP terminology we do not use the term "functions," but rather "nonterminals," and in GGP systems terminals and nonterminals refer to the symbols of the grammar.

The motivation for combining grammars and GP is twofold [59]. First, it allows the user to incorporate prior knowledge about the problem domain to help guide the GP search. Secondly, it guarantees the closure property through the definition of grammar production rules.

GGP has been used in a wide variety of applications. One of its first application domains was to develop the topology of artificial neural networks [31, 35]. It was also used in symbolic function regression [40, 52, 59] to take into account the di-

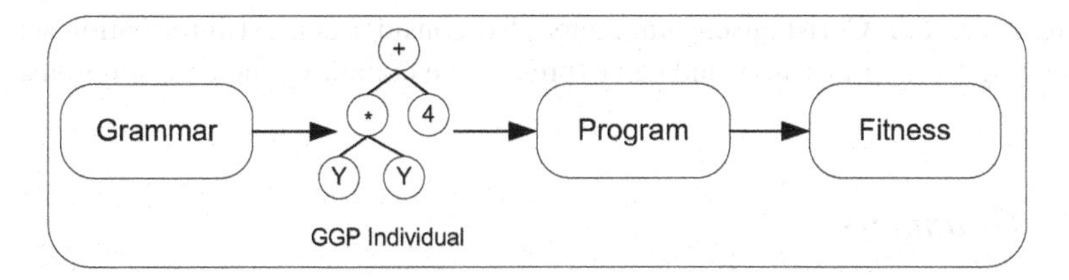

Fig. 3.8 GGP scheme with solution-encoding individual representation

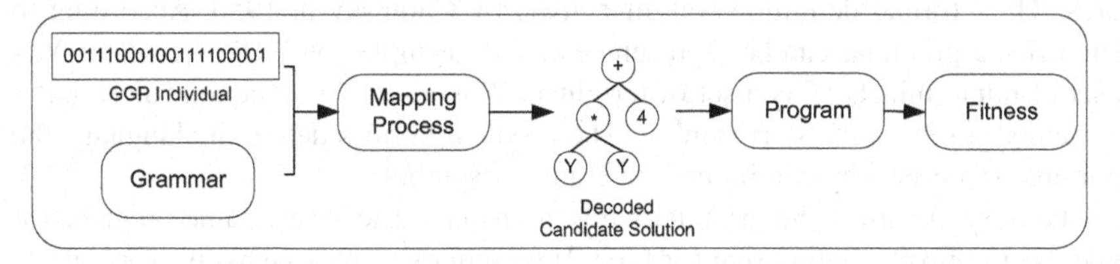

Fig. 3.9 GGP scheme with production-rule-sequence-encoding individual representation

mension of variables when evolving physical laws [67], to perform the clustering task of data mining [17], and to evolve rule sets [58, 75, 80].

GGP systems can be divided into different classes according to two different criteria:

1. The representation used by the GGP individuals.
2. The type of grammar on which the GGP system is based.

Considering the representation of the GGP individuals, GGP systems follow two different approaches: solution-encoding individual and production-rule-sequence-encoding individual, represented in Figs. 3.8 and 3.9, respectively. In the first approach (Fig. 3.8), there is no difference between the individuals' *genotype* (encoded candidate solution representation manipulated by the genetic operators) and *phenotype* (decoded candidate solution from the user's point of view). An individual is represented by a tree, which corresponds to a derivation tree produced when following the production rules of the grammar. The main characteristics of this approach and some GGP systems based on it will be presented in Section 3.6.2.

The second GGP approach (Fig. 3.9) differs from the first because it uses a mapping between the individual genotype and phenotype (the search and solution space). In this approach, the individuals are represented by a linear genome (usually a binary string or an array of integers), which is generated independently of the grammar. When evaluating the individuals, a genotype-phenotype mapping is performed, and the genetic material is used to select appropriate production rules from the grammar, as detailed in Section 3.6.3.

Regarding the types of grammar used to guide the GP, the most popular are the context-free grammars [1]. However, after the popularization of systems combining grammars and GP, work has been done using logic grammars [80], attribute grammars [30], tree-adjunct and tree-adjoining [38] grammars, and Christiansen grammars [60]. While context-free grammars are used to restrict the syntax of the programs generated, Christiansen grammars also consider context information while generating trees (programs), and can express more complex candidate solutions.

3.6.1 Grammars

Grammars [1] are simple mechanisms capable of representing very complex structures. Their formal definition was first given by Chomsky in 1950. According to Chomsky, a grammar can be represented by a four-tuple $\{N, T, P, S\}$, where N is a set of nonterminals, T is a set of terminals, P is a set of production rules, and S (a member of N) is the start symbol. The production rules define the language the grammar represents by combining the grammar symbols.

Chomsky classified the grammars into four main categories, named regular (or type 3) grammars, context-free (or type-2) grammars, context-sensitive (or type 1) grammars and phrase-structure (or type 0) grammars [14]. Regular grammars are the most restrictive and also the simplest grammars, while phrase-structure grammars are the most general grammars, and also very complex. The differences between these classes of grammars are determined by the structure of the production rules they might have. For example, in regular grammars, the production rules can be presented in only two forms: a single nonterminal symbol produces a single terminal symbol; or a single nonterminal symbol produces a single nonterminal symbol followed by a terminal symbol. In contrast, in context-free grammars, a production rule is defined by a nonterminal followed by a string of terminals and/or nonterminals. Grammars belonging to the types 0 and 1 are relatively little used in real-world applications, mainly because their implementation and parsing processes are very complicated.

In this section we are especially interested in context-free grammars (CFGs), which is the type of grammar used by the GGP system proposed in Chapter 5. CFGs are the class of grammars most commonly used with genetic programming. Figure 3.10 shows an example of a CFG that produces simple expressions by combining the operators $+$ and $-$, the variables x and y, and the numbers 2 and 4.

The grammar in Fig. 3.10 is described using the Backus Naur Form (BNF) [56]. When using the BNF notation, production rules have the form <expr> ::= <expr><op><expr>, and symbols wrapped in "<>" represent the nonterminals of the grammar. Three special symbols might be used for writing the production rules in BNF: "|","[ ]", and "( )". "|" represents a choice, as in <var> :: $=x \mid y$, where *<var>* generates the symbol x or y. "[ ]" wraps an optional symbol, which may or may not be generated when applying the rule. "( )"is used to group a set of choices together, as in $x ::= k(y|z)$, where x generates k followed by y or z.

The application of a production rule from $p \in P$ to some nonterminal $n \in N$ is called a derivation step, and it is represented by the symbol "$\Rightarrow$". Figure 3.10 shows the derivation steps necessary to produce the expression $x+2$. These derivation steps can be graphically represented by the derivation tree also presented in Fig. 3.10.

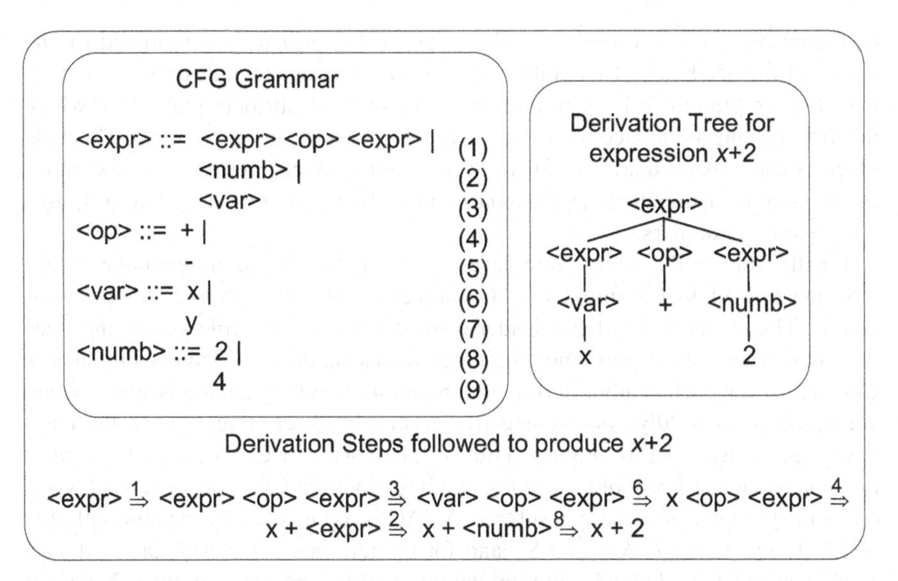

Fig. 3.10 Context-free grammar to create a simple expression, showing an example of a derivation tree and the list of derivation steps followed to generate the expression $x + 2$

Most of the GP systems described in the next subsections are based on a CFG. This is because CFGs are very good methods to enforce the production of syntactically correct solutions. However, in some problems, information about context is essential, and this kind of information cannot be provided by a CFG. Consider, for example, a grammar which describes a language of strings having the same numbers of characters "a," "b," and "c," i.e, L = $a^n b^n c^n$. This language is context-sensitive, because we need to have information about how many characters "a" have been produced in order to produce the same number of "b" and "c" characters.

The language L = $a^n b^n c^n$ cannot be generated by a CFG, but it can be produced by a context-sensitive grammar. As pointed out before, a context-sensitive grammar is complex to implement and difficult to parse. Thus, in cases such as this, instead of opting for a context-sensitive grammar, we can adopt extensions for CFGs that can deal with context. In the case of GGP systems, these extensions include the use of logic grammars (based on definite clause grammars) [80], attribute grammars [35], and Christiansen grammars [60].

Logic grammars are generalizations of a CFG, where the symbols of the grammar (terminals or nonterminals) can include arguments. Arguments can be any term in the grammar, where terms can be a logical variable, a function, or a constant, and are used to enforce context-dependency [81].

Attribute grammars are an extension of CFG where each symbol of the CFG can be associated with one or more attributes. Hence, when generating a derivation tree, each node in the tree is associated with a set of attributes. The values of the attributes in the derivation tree are determined through two kinds of evaluations. In the first case, the value of an attribute is inherited from its parent node (inherited attribute). In the second case, the value of an attribute is determined by the values of the attributes of its child node's attribute values (synthesized attributes). Christiansen grammars [12], in turn, are extensions of attribute grammars, where the first attribute associated with every symbol is a Christiansen grammar. This type of grammar is particularly useful to account for semantic restrictions. Examples of GP systems using attribute grammars can be found in [13, 35], while [60] uses Christiansen grammars.

Finally, some GP systems were developed using tree-adjoining grammar (TAG) [38] and tree-adjunct grammars (TAGs implemented with only the adjoining operator). These grammars also consider context information while generating trees (programs), and can express more complex representations. A TAG is similar to a CFG. However, each symbol (terminal/nonterminal) of the grammar is a tree. TAGs are classified as mildly context-sensitive grammars. According to [32], the main advantage of using TAGs and tree-adjunct grammars over CFGs is that the former implement a natural way of preserving building blocks in GP systems, as each symbol in the grammar is actually a subtree. A TAG is represented by the quintuple $\{T, NT, I, A, S\}$, where T, NT, and S stand for the terminals, nonterminals, and start symbol, as in CFGs; I and A represent the set of initial and auxiliary trees. Note that TAGs do not have production rules, and the derivation trees are created by two other trees (belonging to I and A) put together using adjoining and substitution operators. A more detailed explanation of TAGs can be found in [38], and examples of GP systems using them can be found in [32, 33].

3.6.2 GGP with Solution-Encoding Individual

This section presents an overview of several GGP systems following the approach described in Fig. 3.8, named solution-encoding individual. All the GGP systems described in this section use a GGP individual that directly encodes the solution for the target problem, and do not require any mapping from the search space (genotype) to the solution space (phenotype). This type of individual representation requires a special procedure to initialize the first population of individuals, and to control crossover and mutation operations.

Figure 3.11 shows an example of a solution-encoding individual using the context-free grammar described in Fig. 3.10. The individual is built through a set of derivation steps, and production rules are applied to the tree until all the leaf nodes are represented by terminals. The solution represented by the GGP individual shown in Fig. 3.11 is obtained by reading the leaf nodes of the tree from left to right ($x+2-4$). This same procedure is used to generate all the individuals in the population. In order to guarantee a variety of individuals in the initial population, the initial population of individuals can be initialized using the traditional ramped-half-and-half procedure, explained in Section 3.5, where trees of different shapes and sizes are produced.

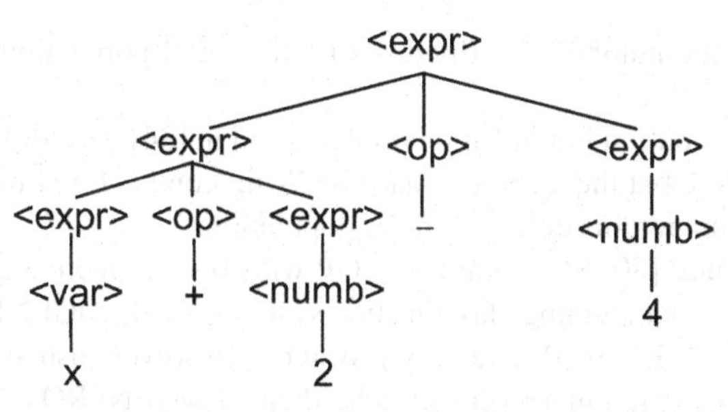

Fig. 3.11 GGP tree representing the expression $x+2-4$

Note that, when choosing a production rule from the grammar to form the individual's derivation tree, the initialization algorithm needs to check whether that particular production rule will be able to reach a terminal symbol in a number of derivation steps smaller than the maximum tree depth permitted.

Crossover and mutation operations are restricted to nonterminals, and different nonterminals might be assigned different crossover and mutation rates. In the case of crossover, a nonterminal N_x is randomly selected from the tree of the first individual I_1. After that, the system searches for the same nonterminal N_x in the tree of individual I_2. If N_x is present in I_2, the subtrees rooted at N_x in individuals I_1 and I_2 are swapped (respecting the maximum individual size parameter). If N_x is not present in I_2, the operation is not performed.

Regarding the mutation operator, a random nonterminal N_x is selected from the derivation tree of the individual, the subtree rooted at N_x is deleted, and a new subtree is created by following the productions of the grammar (starting from N_x).

The population initialization, individual representation, and crossover and mutation operations just described were introduced by Whigham [77], but of course other methods can be used, as long as they respect the grammar's structure. In his work, Whigham describes grammars to solve the six-multiplexer problem and a classification problem named "greater glider density" [78]. For the latter problem, Whigham defines a set of grammars that supports *if-then-else* statements, and builds programs that can be read as a set of rules.

Furthermore, one of the most interesting features of the system created by Whigham is that the grammar is adaptable. At each generation, an analysis of the fittest individuals in the population is performed, and the grammar's production rules are modified according to the results of this evaluation.

Following Whigham, many other researchers developed GGP systems with the solution-encoding individual representation. Horner [34] implemented a C++ GP library able to deal with solution-encoding individuals, which he named GPK (Genetic Programming Kernel). GPK's implementation presents similar features to those of the system proposed in [77]. However, its initialization process is much more complex. It requires the calculation of the search-space size, and the definition of probabilities of individuals of a specific size s (s varying from 0 to the maximum tree depth) to appear in an equally distributed population. These probabilities are later used to set the number of individuals in the initial population, which should have size s.

Indeed, the population initialization process used in [34] was too complicated, as pointed out by [63], but the technique used by Whigham [77] is simple and elegant, and successfully maintains diversity in the population.

Wong and Leung [80, 81] combined GGP with ILP (Inductive Logic Programming) to produce a data mining classification system called LOGENPRO (The Logical grammar based Genetic Programming system). However, instead of a CFG, they worked with a logic grammar to create individuals. LOGENPRO can produce both decision trees and rule sets. Here we are going to describe how it was used to evolve classification rule sets.

When evolving rule sets, LOGENPRO individuals (derivation trees extracted from a logic grammar) represent classification rules. LOGENPRO follows the Michigan approach, where each individual represents a single candidate rule, unlike the Pittsburgh approach, where an individual represents a set of rules [27]. The first population of the GGP system implemented by LOGENPRO is formed by randomly generated individuals, or by individuals extracted from other rule learning systems. The first population can also contain individuals created by the user and given to the system.

LOGENPRO uses crossover and mutation operators in order to produce new valid individuals. The trees representing individuals might present some "frozen" nodes, which cannot be swapped during crossover operations. Crossover swaps subtrees rooted at two nonterminals to produce a single individual, and mutation recreates a subtree rooted at a random nonterminal according to the grammar. Reproduction is not used, but a third operator called drop condition (which generalizes rules) was implemented. The fitness function is based on the support-confidence framework typically used to evaluate association rules in data mining [79].

LOGENPRO also has a mechanism called token competition, which aims to maintain population diversity. It considers each instance in the training set as a token, and the individuals compete for it. The strongest individuals in the population get to collect tokens first. After token competition, the fitness of the individuals is modified according to the number of tokens it collected. Individuals that do not collect any tokens are considered "redundant," and are replaced by new individuals. The individuals (rules) with the best fitness are passed to the next generation without any modification, by elitism. The evaluation of LOGENPRO in benchmarking datasets from the well-known UCI dataset repository [57] and in medical application domains showed results competitive with other rule induction algorithms.

With the same purposes of Wong, Tsakonas et al. [75] proposed to evolve rule sets for two medical domains using GP systems with CFGs. They proposed two systems: one to evolve crisp rules and another to evolve fuzzy rules. Again, the population initialization and evolution operations are the conventional ones in GGP systems with solution-encoding individuals, as proposed by Whigham. In these systems, each GGP individual is a rule list for a specific class. Hence, in a system with C classes, the GGP system is executed C times, in each of them trying to find a rule list which separates the ith class ($i = 1, \dots, C$) from the other $C - 1$ classes.

The fitness of the system to evolve crisp rules is a weighted function that takes into account a correlation measure based on the statistics gathered from the training set and the size of the GGP trees (the size of the rule sets). In the case of the system with fuzzy rules, the simplicity of the rule sets was ignored, as the authors concluded that in the way the grammar was defined, large individuals could still be easily interpreted.

The system was compared with C4.5 [66], with a boosting algorithm, and with a standard GP for symbolic regression in two medical domains. The results showed that the proposed methods found very comprehensible rules (according to the opinion of some experts in the areas), with a classification accuracy competitive with the other methods.

Falco et al. [17] also implemented a GGP system to comply with a data mining task, but instead of classification they worked with clustering. They defined a new way of representing clusters via logical formulas, and used a GGP system to search for these formulas. Comparisons with well-known clustering algorithms showed that GGP is a successful method to create clusters.

McConaghy and Gielen [52] proposed CAFFEINE (CAnonical Form Function Expressions IN Evolution). CAFFEINE makes use of canonical form functions to evolve human-interpretable expressions for symbolic regression problems. These functions are implemented via a CFG. In [52], CAFFEINE was tested using both a GGP system with solution-encoding individuals and with production-rule-sequence-encoding individuals. Both implementations were evaluated in an application of knowledge extraction in analog circuit design, and in this particular problem the GGP system using solution-encoding individuals performed better than the GGP system using production-rule-sequence-encoding individuals.

From all the systems described so far, Wong and Leung [81] is the only one which does not use CFGs. However, other attempts to work with grammars different from the context-free ones were made. Hussain and Browse [35], for instance, proposed the use of an attribute grammar to evolve the topology of artificial neural networks. They used attribute grammars to represent the topology of neural networks in many different levels, in a system they called NGAGE (Network Generating Attribute Grammar Encoding). In NGAGE, the GGP individuals are represented by derivation trees, where each of the leaf nodes represents a neuron in the neural net. The attributes associated with a leaf node define which of the other leaf nodes will be used to form its inputs and outputs. The internal nodes of the derivation tree define the structure of the neural net.

3.6.3 GGP with Production-Rule-Sequence-Encoding Individual

This section presents an overview of several Grammar-based Genetic Programming (GGP) systems following the production-rule-sequence-encoding individual approach illustrated in Fig. 3.9. Note that, as the evolutionary algorithms described in this section use a linear genome that usually does not contain operators or functions, some authors named them as genetic algorithms (GAs), despite the fact that they are being used to evolve programs. Although indeed the GA term is appropriate to describe the evolutionary algorithm itself, the whole system is actually composed of the GA plus the grammar, where the latter contains operators or functions. Hence, we will refer to such systems as a whole as GP (rather than GA) systems, and more precisely as GGP systems. Regardless of the name these systems receive, here we focus on the role of the grammar in them, and how the genotype-phenotype mapping process is implemented.

All the GGP systems based on this approach present two common elements that distinguish them from the GGP systems described in Section 3.6.2:

1. Their individuals are represented by a linear chromosome, which can be fixed or variable in length, and the chromosome contains a string of bits or integers.
2. The search space is distinct from the solution space, and each gene in a chromosome usually points to a production rule in the grammar. Hence, a mapping process is needed to generate solutions by reading the individuals' chromosomes.

As this approach works with linear genomes, usually no restrictions are needed in the way the initial population is created, or in the way crossover and mutation operators are applied. Hence, these operations are simpler than their tree-based counterparts.

Banzhaf and Keller [6, 40] described one of the first systems using linear genomes and grammars. In their system, the fixed-length binary genotype was divided into codons composed of a predetermined number of bits. Each codon was mapped to a symbol of the output language, and different codons could lead to the same symbol (redundant genetic code). They argued that redundant genetic code leads to more diverse populations, as in natural systems. However, the role of the grammar in this system was to act as a correction mechanism in the case of invalid outputs being generated when combining the output symbols.

Following the same principles of the work described in [6], Paterson and Livesey [62] proposed GADS (Genetic Algorithm for Deriving Software). GADS uses a fixed-length integer array to represent an individual, where each gene in the individual chromosome (an integer number) points to a production rule in the grammar. The main difference between the works presented in [6] and [62] is the role of the grammar in the system. While in the former the grammar is just a correction mechanism, in the latter the grammar is actually responsible for the generation of solutions.

In [62], after the population of linear individuals is initialized, a mapping process between the genotype and the phenotype of an individual starts by automatically inserting the start symbol in the root of the individual's tree and reading the genes in the chromosome from left to right. The production rule indicated by a gene may or may not be used. It is used if the nonterminal in the left-hand side of the production rule is present in the individual tree. Otherwise, it is ignored, and the next gene in the chromosome is read. The mapping process finishes when all the genes in the chromosome are read. At this point, if there are still nonterminals in the derivation tree, they are replaced by their default values. Note that each nonterminal of the grammar is associated with a default production rule.

After the works of Keller and Banzhaf [6, 40] and Paterson and Livesey [62], some other authors proposed small modifications to mitigate the drawbacks of the previous approaches. For instance, Freeman [25] improved the control of the proliferation of introns in [62] by introducing simple modifications into the mapping process. However, in 1998, Ryan et al. [69] introduced what they called grammatical evolution (GE) [59]. In contrast with the previous approaches, GE uses a variable-length binary string genome divided into codons of eight bits. The codons are used to select appropriate production rules from the grammar, and the system has mechanisms to prevent the selection of invalid rules, preventing the appearance of introns.

The mapping process works by translating a codon into an integer value, and then dividing this value by the number of available production rules for the nonterminal being extended. The remainder of this division is then used to select a rule from the set of available rules. For each nonterminal, its production rules are numbered from 0 to n. Consider for example the grammar presented in Fig. 3.10. The nonterminal <*op*> produces the terminals $+$ or $-$. Assume the algorithm is reading the codon 00010101. This codon is translated into the integer 21, which is divided by 2 (number of production rules available having <*op*> in their antecedent). The remainder of this division is 1, and so the production rule which generates the "−" terminal is chosen. In the case that the individual runs out of codons and there are still nonterminals to be expanded, an operation called wrap starts reading the genotype of the individual again and reuses its codons.

In terms of evolution, GE uses a steady-state approach, rather than a generational approach, and the offspring replace individuals in the same population. The genetic operators are the simple ones used in a standard genetic algorithm, including one-point crossover and point mutation. A third operator, named codon duplication operator, is also introduced. In [58] we can find an example of a GE system used to evolve market index trading rules. The system evolved a set of fuzzy *if-then* rules which outperformed the baseline buy-and-hold strategy.

The GE approach has been applied to a variety of problems in the past years, and new variations of this system have emerged. CHORUS [3], for instance, implements a new version of the system with a position-independent representation of individuals. The main difference between CHORUS and GE lies in the mapping process. In CHORUS, instead of numbering the production rules of each nonterminal independently, all the productions of the grammar are numbered together (similarly to the system used in [62]). As a result, during the mapping process, the integer extracted from the chromosome is divided by the total number of rules in the grammar. In this way, one gene will always represent the same rule in the grammar, regardless of its position in the chromosome.

CHORUS also works with a concentration table, which helps select rules during the mapping process. The concentration table has a column for each rule in the grammar, and is used to keep track of the count (or concentration) of the rules. Every time a gene is read, it increases the concentration of the rule it represents. The algorithm stops reading the genes when the concentration of one of the applicable rules is greater than that of the others, and marks the position of the gene to be read later.

Apart from CHORUS, another work extends the standard GE approach to work with attribute grammars [13]. Attribute grammars, as explained before, take into account context-sensitive information, and have a considerably larger expressiveness power than CFGs. This particular work implements a GE system with attribute grammars to solve the knapsack problem – a well-known combinatorial optimization problem. For more details the reader is referred to [13].

3.7 Summary

This chapter presented the basic concepts of evolutionary algorithms, and introduced their four main components: individual representation, fitness function, selection methods, and genetic operators. It also discussed the differences between genetic algorithms and genetic programming, emphasizing that the main difference between these two types of evolutionary algorithms is not their different individual representation (linear strings versus trees), but rather the fundamental nature of the candidate solution produced by each type of algorithm. In particular, genetic algorithms typically find solutions for a single instance of a problem, while in principle GP aims or should aim at finding programs or algorithms capable of solving any instance of the target type of problem.

After a somewhat detailed discussion about standard genetic programming systems, we presented one of its most powerful variations: grammar-based genetic programming (GGP). We stressed GGP systems' ability to introduce background knowledge about the problem into the system, and their ability to easily cope with the closure problem.

We also briefly reviewed some systems that use GGP systems for solving a variety of problems, and classified them into two main types according to their individual representation: GGP with solution-encoding individual and GGP with production-rule-sequence-encoding individual. While there is a differentiation between genotype and phenotype in the latter, in the former both the genotype and the phenotype of an individual are the same. A GGP system with solution-encoding individual representation will be used in Chapter 5 to automatically evolve rule induction algorithms.

References

1. Aho, A.V., Sethi, R., Ullman, J.D.: Compilers: Principles, Techniques and Tools, 1st edn. Addison-Wesley (1986)
2. Angeline, P.J.: Subtree crossover causes bloat. In: J.R. Koza, W. Banzhaf, K. Chellapilla, K. Deb, M. Dorigo, D.B. Fogel, M.H. Garzon, D.E. Goldberg, H. Iba, R. Riolo (eds.) Proc. of the 3rd Annual Conf. on Genetic Programming (GP-98), pp. 745–752. Morgan Kaufmann (1998)

3. Azad, R.M.A.: A position independent representation for evolutionary automatic programming algorithms – the Chorus system. Ph.D. thesis, University of Limerick, Ireland (2003)
4. Baeck, T., Fogel, D.B., Michalewicz, Z.: Evolutionary Computation 1 : Basic Algorithms and Operators. Institute of Physics Publishing (2000)
5. Bandyopadhyay, S., Pal, S.: Classification and Learning Using Genetic Algorithms. Springer (2007)
6. Banzhaf, W.: Genotype-phenotype-mapping and neutral variation – a case study in genetic programming. In: Y. Davidor, H. Schwefel, R. Männer (eds.) Parallel Problem Solving from Nature III, *LNCS*, vol. 866, pp. 322–332. Springer-Verlag, Jerusalem (1994)
7. Banzhaf, W., Langdon, W.: Some considerations on the reason for bloat. Genetic Programming and Evolvable Machines **3**, 81–91 (2002)
8. Banzhaf, W., Nordin, P., Keller, R.E., Francone, F.D.: Genetic Programming – An Introduction; On the Automatic Evolution of Computer Programs and its Applications. Morgan Kaufmann (1998)
9. Basgalupp, M., Barros, R., Carvalho, A., Freitas, A., Ruiz, D.: Legal-tree: a lexicographic multi-objective genetic algorithm for decision tree induction. In: Proc. of 24th Annual ACM Symposium on Applied Computing (SAC 2009), Hawaii, USA, pp. 1085–1090 (2009)
10. Blickle, T.: Tournament selection. In: T. Back, D. Fogel, T. Michalewicz (eds.) Evolutionary Computation 1: Basic Algorithms and Operators, pp. 181–186. Institute of Physics Publishing (2000)
11. Breiman, L., Friedman, J., Olshen, R., Stone, C.: Classification and Regression Trees. Wadsworth (1984)
12. Christiansen, H.: A survey of adaptable grammars. SIGPLAN Not. **25**(11), 35–44 (1990)
13. Cleary, R.: Extending grammar evolution with attribute grammars: an application to knapsack problems. Master's thesis, University of Limerick, Canberra, Australia (2005)
14. Cleaveland, J., Uzgalis, R.C.: Grammars for Programming Languages. Elsevier Computer Science Library, New York, USA (1977)
15. Coello, C.A.C., Lamont, G. (eds.): Multi-Objective Algorithms for Attribute Selection in Data Mining. World Scientific (2004)
16. Coello, C.C., Veldhuizen, D.V., Lamont, G.: Algorithms for Solving Multi-Objective Problems. Kluwer Academic Publishers, New York, USA (2002)
17. De Falco, I., Tarantino, E., Cioppa, A.D., Fontanella, F.: A novel grammar-based genetic programming approach to clustering. In: Proc. of the 2005 ACM Symposium on Applied Computing (SAC-05), pp. 928–932. ACM Press, New York, NY, USA (2005)
18. Deb, K.: Introduction to selection. In: T. Back, D. Fogel, T. Michalewicz (eds.) Evolutionary Computation 1: Basic Algorithms and Operators, pp. 166–171. Institute of Physics Publishing (2000)
19. Deb, K.: Multi-Objective Optimization Using Evolutionary Algorithms. Wiley Interscience series in Systems and Optimization, Berlin (2001)
20. Domingos, P.: Occam's two razors: the sharp and the blunt. In: Proc. of the 4th Int. Conf. on Knowledge discovery and data mining (KDD-98), pp. 37–43 (1998)
21. Eiben, A.E., Smith, J.E.: Introduction to Evolutionary Computation. Springer-Verlag (2003)
22. Emmanouilidis, C., Hunter, A., MacIntyre, J.: A multiobjective evolutionary setting for feature selection and a commonality-based crossover operator. In: Proc. of the Congress on Evolutionary Computation (CEC-00), pp. 309–316. IEEE Press (2000)
23. Eshelman, L., Caruana, R., Schaffer, J.: Biases in the crossover landscape. In: Proc. of the Int. Conf. on Genetic Algorithms (ICGA-89), pp. 10–19 (1989)
24. Falkenauer, E.: The worth of the uniform. In: Proc. of the Congress on Evolutionary Computation(CEC-99), pp. 1776–782. IEEE (1999)
25. Freeman, J.J.: A linear representation for GP using context free grammars. In: J.R. Koza, W. Banzhaf, K. Chellapilla, K. Deb, M. Dorigo, D.B. Fogel, M.H. Garzon, D.E. Goldberg, H. Iba, R. Riolo (eds.) Proc. of the 3rd Annual Conf. on Genetic Programming (GP-98), pp. 72–77. Morgan Kaufmann (1998)
26. Freitas, A.: A review of evolutionary algorithms for data mining. In: Soft Computing for Knowledge Discovery and Data Mining, pp. 79–111. Springer (2007)

27. Freitas, A.A.: Data Mining and Knowledge Discovery with Evolutionary Algorithms. Springer-Verlag (2002)
28. Freitas, A.A.: A critical review of multi-objective optimization in data mining: a position paper. ACM SIGKDD Explorations Newsletter **6**(2), 77–86 (2004)
29. Goldberg, D.E.: Genetic Algorithms in Search, Optimization, and Machine Learning. Addison-Wesley, Reading, MA (1989)
30. Goos, G., Hartmanis, J. (eds.): Attribute Grammar: Definition, Systems and Bibliography. Lecture Notes in Computer Science (1988)
31. Gruau, F.: On using syntactic constraints with genetic programming. In: P.J. Angeline, K.E. Kinnear Jr., (eds.) Advances in Genetic Programming 2, chap. 19, pp. 377–394. MIT Press, Cambridge, MA, USA (1996)
32. Hoai, N.X., McKay, R.I., Abbass, H.A.: Tree adjoining grammars, language bias, and genetic programming. In: C. Ryan, T. Soule, M. Keijzer, E. Tsang, R. Poli, E. Costa (eds.) Proc. of the 6th European Conf. on Genetic Programming (EuroGP-03), vol. 2610, pp. 335–344. Springer-Verlag (2003)
33. Hoai, N.X., McKay, R.I., Essam, D.: Some experimental results with tree adjunct grammar guided genetic programming. In: J.A. Foster, E. Lutton, J. Miller, C. Ryan, A.G.B. Tettamanzi (eds.) Proc. of the 5th European Conf. on Genetic Programming (EuroGP-02), *LNCS*, vol. 2278, pp. 228–237. Springer-Verlag (2002)
34. Horner, H.: A C++ class library for genetic programming: the Vienna University of Economics genetic programming kernel. citeseer, citeseer.nj.nec.com/horner96class.html (1996)
35. Hussain, T., Browse, R.: Network generating attribute grammar encoding. In: Proc. of IEEE Int. Joint Conf. on Neural Networks, pp. 431–436 (1998)
36. Hutt, B., Warwick, K.: Synapsing variable-length crossover: meaningful crossover for variable-length genomes. IEEE Transactions on Evolutionary Computation **11**(1), 118–131 (2007)
37. Jin, Y. (ed.): Multi-Objective Machine Learning. Springer, Berlin (2006)
38. Joshi, A.K., Schabes, Y.: Tree-adjoining grammars. In: G. Rozenberg, A. Salomaa (eds.) Handbook of Formal Languages, vol. 3, pp. 69–124. Springer, Berlin, New York (1997)
39. Kaufmann, K., Michalski, R.: Learning from inconsistent and noisy data: the AQ18 approach. In: 11th Int. Symposium on Foundations of Intelligent Systems (ISMIS-99), pp. 411–419. Springer (1999)
40. Keller, R.E., Banzhaf, W.: Genetic programming using genotype-phenotype mapping from linear genomes into linear phenotypes. In: J.R. Koza, D.E. Goldberg, D.B. Fogel, R.L. Riolo (eds.) Proc. of the 1st Annual Conf. on Genetic Programming (GP-96), pp. 116–122. MIT Press, Stanford University, CA, USA (1996)
41. Kishore, J., Patnaik, L., Mani, V., Agrawal, V.: Application of genetic programming for multicategory pattern classification. IEEE Transactions on Evolutionary Computation **4**(3), 242–258 (2000)
42. Koza, J.R.: Genetic Programming: on the programming of computers by the means of natural selection. The MIT Press, Massachusetts (1992)
43. Koza, J.R., Keane, M.A., Streeter, M.J., Mydlowec, W., Yu, J., Lanza, G.: Genetic Programming IV: Routine Human-Competitive Machine Intelligence. Kluwer Academic Publishers (2003)
44. Kudo, M., Sklansky, J.: Comparison of algorithms that select features for pattern classifiers. Pattern Recognition **33**, 25–41 (2000)
45. Langdon, W.: Fitness causes bloat in variable size representations. In: Proc. of the Workshop on Evolutionary Computation with Variable Size Representation, (ICGA-97) (1997)
46. Langdon, W., Soule, T., Poli, R., Foster, J.: The evolution of size and shape. In: L. Spector, W. Langdon, U.M. O'Reilly, P. Angeline (eds.) Advances in Genetic Programming, Vol. 3, pp. 184–191. MIT (1999)
47. Langdon, W.B., Poli, R.: Fitness causes bloat: mutation. In: Proc. of the 1st European Workshop on Genetic Programming (EuroGP-98) (1998)

48. Li, Q., Wang, W., Han, S., Li, J.: Evolving classifier ensemble with gene expression programming. In: Proc. 3rd Int. Conf. on Natural Computation (ICNC-07). IEEE Computer Society (2007)
49. Liu, H., Motoda, H. (eds.): Computational Methods of Feature Selection. Chapman and Hall/CRC, London (2008)
50. Llorà, X., Garrell, J.M.: Prototype induction and attribute selection via evolutionary algorithms. Intelligent Data Analysis **7**(3), 193–208 (2003)
51. Majeed, H., Ryan, C.: Using context-aware crossover to improve the performance of GP. In: Proc. of the Genetic and Evolutionary Computation Conf. (GECCO-06), pp. 847–854. ACM Press (2006)
52. McConaghy, T., Gielen, G.: Canonical form functions as a simple means for genetic programming to evolve human-interpretable functions. In: Proc. of the Genetic and Evolutionary Computation Conf. (GECCO-06), pp. 855–862. ACM Press, New York, NY, USA (2006)
53. Michalewicz, Z.: Genetic Algorithms + Data Structures = Evolution Programs, 3 edn. Springer (1996)
54. Montana, D.J.: Strongly typed genetic programming. Evolutionary Computation **3**(2), 199–230 (1995)
55. Muni, D., Pal, N., Das, J.: A novel approach to design classifier using genetic programming. IEEE Transactions on Evolutionary Computation **8**(2), 183–196 (2004)
56. Naur, P.: Revised report on the algorithmic language ALGOL-60. Communications ACM **6**(1), 1–17 (1963)
57. Newman, D.J., Hettich, S., Blake, C.L., Merz, C.J.: UCI Repository of machine learning databases. University of California, Irvine, http://www.ics.uci.edu/~mlearn/MLRepository.html (1998)
58. O'Neill, M., Brabazon, A., Ryan, C., Collins, J.: Evolving market index trading rules using grammatical evolution. In: E.J.W. Boers, S. Cagnoni, J. Gottlieb, E. Hart, P.L. Lanzi, G.R. Raidl, R.E. Smith, H. Tijink (eds.) Applications of Evolutionary Computing, *LNCS*, vol. 2037, pp. 343–352. Springer-Verlag (2001)
59. O'Neill, M., Ryan, C.: Grammatical Evolution : Evolutionary Automatic Programming in an Arbitrary Language. Morgan Kaufmann (2003)
60. Ortega, A., de la Cruz, M., Alfonseca, M.: Christiansen grammar evolution: grammatical evolution with semantics. IEEE Transactions on Evolutionary Computation **11**(1), 77–90 (2007)
61. Pappa, G.L., Freitas, A.A., Kaestner, C.A.A.: Multi-objective algorithms for attribute selection in data mining. In: C.A.C. Coello, G. Lamont (eds.) Applications of Multi-Objective Evolutionary Algorithms, pp. 603–626. World Scientific (2004)
62. Paterson, N.R., Livesey, M.: Distinguishing genotype and phenotype in genetic programming. In: J.R. Koza (ed.) Late Breaking Papers of the 1996 Annual Conf. on Genetic Programming, pp. 141–150. Stanford Bookstore, Stanford University, CA, USA (1996)
63. Paterson, N.R., Livesey, M.: Evolving caching algorithms in C by genetic programming. In: J.R. Koza, K. Deb, M. Dorigo, D.B. Fogel, M. Garzon, H. Iba, R.L. Riolo (eds.) Proc. of the 2nd Annual Conf. on Genetic Programming (GP-97), pp. 262–267. Morgan Kaufmann, Stanford University, CA, USA (1997)
64. Philemotte, C., Bersini, H.: A gestalt genetic algorithm: less details for better search. In: Proc.of the Genetic and Evolutionary Computation Conf. (GECCO-07), pp. 1328–1334. ACM Press (2007)
65. Poli, R., Langdon, W., McPhee, N.: A Field Guide to Genetic Programming. Freely available at http://www.gp-guide.org.uk (2008)
66. Quinlan, J.R.: C4.5: programs for machine learning. Morgan Kaufmann (1993)
67. Ratle, A., Sebag, M.: Genetic programming and domain knowledge: beyond the limitations of grammar-guided machine discovery. In: M. Schoenauer, K. Deb, G. Rudolph, X. Yao, E. Lutton, J.J. Merelo, H. Schwefel (eds.) Proc. of the 6th Int. Conf. on Parallel Problem Solving from Nature (PPSN-00), pp. 211–220. Springer Verlag (2000)
68. Rothlauf, F.: Representations for Genetic and Evolutionary Algorithms (2nd Ed.). Springer, Berlin (2006)

69. Ryan, C., Collins, J.J., O'Neill, M.: Grammatical evolution: evolving programs for an arbitrary language. In: W. Banzhaf, R. Poli, M. Schoenauer, T.C. Fogarty (eds.) Proc. of the 1st European Workshop on Genetic Programming, *Lecture Notes in Computer Science*, vol. 1391, pp. 83–95. Springer-Verlag, Paris (1998)
70. Seo, D., Moon, B.: Voronoi quantized crossover for traveling salesman problem. In: Proc. of the Genetic and Evolutionary Computation Conf. (GECCO-02), pp. 544–552. Morgan Kaufmann (2002)
71. da Silva, S.G.O.: Controlling bloat: individual and population based approaches in genetic programming. Ph.D. thesis, University of Coimbra, Departamento de Engenharia Informatica, Coimbra, Portugal (2008)
72. Syswerda, G.: Uniform crossover in genetic algorithms. In: Proc. of the 2nd Int. Conf. on Genetic Algorithms (ICGA-89), pp. 2–9 (1989)
73. Thie, C., Giraud-Carrier, C.: Learning concept descriptions with typed evolutionary programming. IEEE Transactions on Knowledge and Data Engineering **17**(12), 1664–1677 (2005)
74. Tsai, H., Yang, J., Kao, C.: Applying genetic algorithms to finding the optimal gene order in displaying the microarray data. In: Proc. of the Genetic and Evolutionary Computation Conf. (GECCO-02), pp. 610–617. Morgan Kaufmann (2002)
75. Tsakonas, A., Dounias, G., Jantzen, J., Axer, H., Bjerregaard, B., von Keyserlingk, D.G.: Evolving rule-based systems in two medical domains using genetic programming. Artificial Intelligence in Medicine **32**(3), 195–216 (2004)
76. Webb, G.I.: Further experimental evidence against the utility of Occam's razor. Journal of Artificial Intelligence Research **4**, 397–417 (1996)
77. Whigham, P.A.: Grammatically-based genetic programming. In: J.P. Rosca (ed.) Proc. of the Workshop on Genetic Programming: From Theory to Real-World Applications, pp. 33–41. Tahoe City, California, USA (1995)
78. Whigham, P.A.: Grammatical bias for evolutionary learning. Ph.D. thesis, School of Computer Science, University College, University of New South Wales, Australian Defence Force Academy, Canberra, Australia (1996)
79. Witten, I.H., Frank, E.: Data Mining: Practical Machine Learning Tools and Techniques with Java Implementations, 2nd edn. Morgan Kaufmann (2005)
80. Wong, M.L.: An adaptive knowledge-acquisition system using generic genetic programming. Expert Systems with Applications **15**(1), 47–58 (1998)
81. Wong, M.L., Leung, K.S.: Data Mining Using Grammar-Based Genetic Programming and Applications. Kluwer, Norwell, MA, USA (2000)
82. Xie, H., Zhang, M., Andreae, P.: An analysis of constructive crossover and selection pressure in genetic programming. In: Proc. of the Genetic and Evolutionary Computation Conf. (GECCO-07), pp. 1739–1746. ACM Press (2007)
83. Zhou, C., Xiao, W., Tirpak, T., Nelson, P.: Evolving accurate and compact classification rules with gene expression programming. IEEE Transactions on Evolutionary Computation **7**(6), 519–531 (2003)

Chapter 4
Genetic Programming for Classification and Algorithm Design

4.1 Introduction

This chapter discusses previous work on the use of Genetic Programming (GP) for solving three different kinds of problems, namely (a) producing a classification model, in which case the GP algorithm is used directly as a classification algorithm, producing a classification model that can be used to predict the classes of new examples; (b) automating, partially, the design of classification algorithms; and (c) automating the design of combinatorial optimization algorithms. Although the latter type of problem is quite different from the classification task of data mining, which is the focus of this book, it is discussed here because the automation of the design of combinatorial optimization algorithms is a research area more developed than the automation of the design of data mining algorithms. Hence, some ideas from the former can be used in the latter.

This chapter is organized as follows. Section 4.2 discusses the differences between a classification model and a classification algorithm, a crucial point for understanding the GP system proposed in Chapter 5. Section 4.3 discusses the use of GP for evolving classification models. Since this book's main contribution is to propose a GP system for automating the design of rule induction algorithms (a specific type of classification algorithm discussed in Chapter 2), this section focuses on evolving models in the form of classification rules or related knowledge representations, such as classification functions or decision trees, that can be easily converted into classification rules.

The next two sections discuss the two works most related to this book's aforementioned main contribution. Section 4.4 reviews previous work proposing a GP system for automating the design of a specific component (namely the rule evaluation function) of a rule induction algorithm, rather than automating the design of a full rule induction algorithm as proposed in this book. Section 4.5 reviews previous work proposing a GP system for automating the design of classification systems in general. That work consists of evolving both dataset-related and algorithm-related aspects of a classification system, rather than focusing on rule induction algorithms,

G.L. Pappa, A.A. Freitas, *Automating the Design of Data Mining Algorithms*,
Natural Computing Series, DOI 10.1007/978-3-642-02541-9_4,

and the GP system proposed in that work manipulates building blocks defined at a coarser-grained level of abstraction by comparison with the finer-grained building blocks manipulated by the GP system proposed in this book.

It should be noted that there are also evolutionary algorithms for evolving components of some types of classification algorithms very different from rule induction algorithms. For instance, evolutionary algorithms have been used to automate the design of the kernel of Support Vector Machines (SVMs) [23] and to automate the design of the learning rules of Artificial Neural Networks (ANNs) [2], sometimes evolving also parameters of the classification model (i.e., the trained neural network) produced by the ANN algorithm, such as the number of neurons in the network and the weights of their interconnections [34, 43]. These works are not further discussed in this book because SVM and ANN algorithms are very different from rule induction algorithms. In particular, those two types of algorithms typically produce "black-box" classification models, which normally cannot be interpreted by the user. In contrast, as explained in Chapter 2, this book focuses on rule induction algorithms, which have the advantage of producing comprehensible ("white-box") classification models, which can be interpreted by the user to try to get some new insight about the underlying data and application domain. In any case, considering that both SVM and ANN are powerful types of classification algorithms with respect to the goal of maximizing predictive accuracy, the development of GP systems for automating the design of SVM and ANN algorithms is also an interesting research area.

Section 4.6 discusses previous work on automating the design of optimization algorithms. It first explains important differences between classification and optimization and reviews the basic idea of "hyper-heuristics" as a type of method for automatically generating heuristics or algorithms. Then it discusses previous work on GP algorithms for evolving heuristics for two types of combinatorial optimization problems – namely the well-known Traveling Salesman Problem and the Satisfiability problem – and for evolving an evolutionary algorithm to be used as an optimization algorithm.

4.2 Classification Models Versus Classification Algorithms

This book focuses on the use of evolutionary algorithms – in particular Genetic Programming (GP) – for automating the design of classification algorithms. It should be noted that this is very different from, and much more challenging than, using evolutionary algorithms for discovering a classification model for a given dataset. In order to understand the crucial difference between these two alternative goals for an evolutionary algorithm, let us recall the basic differences between the concepts of classification model and classification algorithm in the context of data mining.

A *classification algorithm* is a computational procedure that receives as input a *training set* of examples, tries to discover a predictive relationship between the predictor attributes and the classes over the examples in the training set, and produces

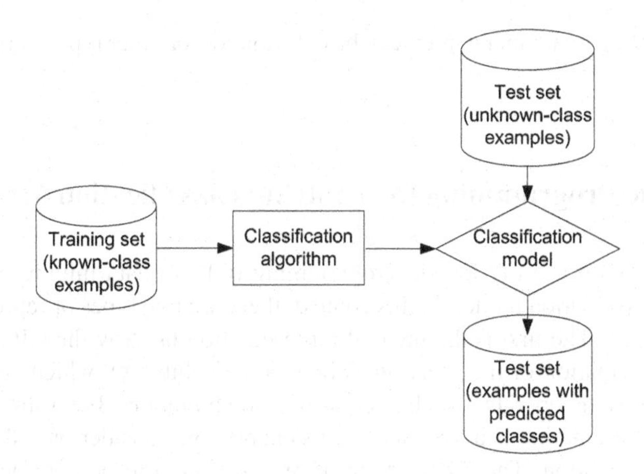

Fig. 4.1 Classification model vs. classification algorithm

as output a *classification model* that expresses the discovered predictive relationship. A *classification model* is a computational procedure that receives as input a *test set* of examples and assigns to each test example a predicted class, based on the values of the example's predictor attributes.

The differences between a classification algorithm and a classification model are clearly shown in Fig. 4.1. Although both are essentially computational procedures, there are very important differences between them. First, they have different types of input: a classification algorithm receives as input the training set (with known-class examples), while a classification model receives as input the test set (with unknown-class examples). Secondly, they have very different types of output: a classification algorithm outputs a classification model, and a classification model outputs predicted classes for the examples in the test set. Thirdly, and the most important point to understand the main contribution of this book, they implement very different types of computational procedures, using very different types of structure and elements, and as a result they have very different types of generality.

Let us elaborate on these points about computational structure and generality. Typically a classification model is not a full-fledged program, but rather has a relatively simple computational structure by comparison with the typically more complex structure of a classification algorithm. This can be clearly observed, for instance, by comparing a decision tree induction algorithm (a type of classification algorithm) like the one shown in Alg. 2.1 with a decision tree (a type of classification model) like the one shown in Fig. 2.5. Turning to different types of generality, note that a classification model has a certain level of generality in the sense that it can be applied to classify any example in the same application domain from which the model was created. In contrast, a classification algorithm has an entirely different level of generality. In principle, we can apply a classification algorithm to many different types of classification datasets belonging to many different types of application domains, e.g., financial, medical, bioinformatics, or other types of application domains.

4.3 Genetic Programming for Evolving Classification Models

The most common use of Genetic Programming (GP) in data mining consists of evolving a classification model. In this context, there are two types of representation to be considered. The first is the internal representation used by the GP algorithm, i.e., the representation of the individuals (candidate solutions), which will be manipulated by genetic operators such as crossover and mutation. By far the most used individual representation is a tree, and in this chapter we consider only this kind of internal representation. The second type of representation to be considered is the knowledge representation from the point of view of the data mining user, i.e., the classification model representation. In this section we discuss the use of an internal tree representation to produce classification models of two different kinds from a data mining point of view, namely classification functions (which can be expressed as *if-then* classification rules) and decision trees.

Before we proceed, there is an important remark to be made about all the GP algorithms discussed in this section. Recall that we argued in Section 3.4 that GP algorithms and other types of evolutionary algorithms such as genetic algorithms are supposed to evolve individuals that represent fundamentally different types of candidate solutions with respect to expressiveness power and generality. Concerning expressiveness power, one would expect that GP individuals (unlike the individuals of genetic algorithms) would represent full programs in the normal sense of the word in computer science, in particular programs containing loops; but GP algorithms producing such programs are difficult to find in the literature, unfortunately.

Concerning the generality of the evolved solutions, one would expect that GP algorithms would evolve very generic programs (algorithms) that, in the context of data mining, could be applied to datasets from different application domains, in the same way that human-designed data mining algorithms (such as the very well-known C4.5 algorithm [33]) can be applied to datasets from different application domains. This is in contrast with other evolutionary algorithms such as genetic algorithms, which are supposed to evolve classification models (rather than classification algorithms) for the application domain and dataset at hand.

All the GP algorithms for evolving classification models discussed in this section fail both the above criteria related to expressiveness power and generality of evolved solutions. None of them evolves programs with loops, and all of them evolve a classification model, and not a classification algorithm. Recall the fundamental distinction between these two types of solution as discussed in Section 4.2.

In this sense, we could perhaps refer to all the evolutionary algorithms discussed in this section as genetic algorithms, and not really GP algorithms, but we will instead use the term GP algorithms to be consistent with the vast literature on this topic. In any case, this point is important because the main contribution of this book is to propose a GP system for automating the design of rule induction algorithms that can be considered a genuine GP system according to the above criteria of expressiveness power and generality: it evolves rule induction algorithms with loops and other typical programming constructs and it evolves very generic rule induction algorithms that can be applied to datasets of any application domain, as will be discussed in detail later.

4.3.1 Evolving Classification Functions or Classification Rules

A popular approach to using GP for evolving a classification model works as follows. The GP terminal set consists of the set of predictor attributes of the data being mined, and possibly some randomly generated constants. The GP function set consists of functions or operators appropriate to manipulate the data types of the attributes (and possibly constants) in the terminal set. A tree-based individual representation is typically used, where the terminals (attributes) are associated with leaf nodes and the functions or operators are associated with internal nodes. A simple example of a conventional GP tree used as a classification model is shown in Fig. 4.2(a).

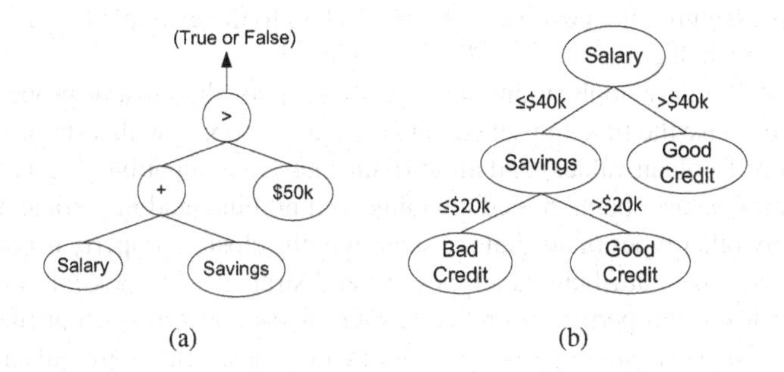

Fig. 4.2 A simple example of a conventional GP tree (a) vs. a decision tree (b)

An individual can then be used as a classification model as follows. For each example (record, data instance) to be classified, the predictor attributes at the leaf nodes of the individual's tree are instantiated with the actual values of the corresponding attributes in that example. Once we apply the functions at the internal (non-leaf) nodes of the individual's tree to the values of the attributes at the leaf nodes, the system computes a value that is output at the root of the individual's tree. This output value is then interpreted in some way as the class assigned to that example. Since the individual as a whole is returning a value which is then used for classification purposes, we will refer to this kind of classification model as a "classification function." This term should not be confused with the functions (or operators) composing the function set of the GP algorithm (typically Boolean or arithmetic functions), which are the basic types of elements used to evolve a classification function. The term classification function is used to refer to the classification model represented by an entire individual (an entire tree). A classification function can also be interpreted as an *if-then* classification rule, as will be discussed later.

In many applications all attributes are numeric and only mathematical functions are used in the function set, so the value output at the root of an individual is a numerical value. This value can be interpreted as a class to be assigned to the current example based on the use of some thresholds to discriminate between classes. In the simple case of binary classification, where the goal is to discriminate between two classes – say, the positive and the negative classes – one can assign the example being classified to the positive class if the output value is greater than or equal to zero, and assign the example to the negative class otherwise. In the case of a multi-class classification problem, a common option is to run the GP algorithm once for each class, i.e., in the ith run, where $i = 1, \ldots, k$ and k is the number of classes, the GP algorithm evolves a classification model to discriminate between class i and all the other classes. Alternatively, one can use a more sophisticated GP algorithm that builds a multi-class classification model in a single run; see, e.g., [37].

If all the attributes are nominal, it is common to use logical operators (e.g., *and*, *or*) in the function set, and in this case the value output at the root node of an individual's tree (after processing the attribute values of an example to be classified) is typically a Boolean value. In a two-class problem, that output value can be interpreted as assigning the positive or negative class to the example being classified, depending on whether the output value is true or false.

Note that in many applications it is easy to satisfy the closure property (Section 3.5.4) because the function set contains operators coping with a single data type (numerical or Boolean values) and all attributes have a compatible data type. However, when the dataset being mined contains both nominal and numerical attributes (and possibly other types of attributes), satisfying the closure property is not so easy. In the case of mixed attribute data types, several strategies are possible in order to satisfy the closure property or to relax it, such as (a) converting all attributes to a single data type (say Boolean) in a preprocessing phase and using only functions coping with that data type in the function set [8, 14, 24]; (b) using some kind of constrained-syntax or strongly typed GP [26, 28]; or (c) using grammar-based GP [42]. For a review of these different approaches, see [18].

As mentioned earlier, a classification function represented in the form of a tree can be interpreted as an *if-then* classification rule. Consider, for instance, the tree shown in Fig. 4.2(a), representing a Boolean classification function. Although the *if-then* structure of a classification rule is not explicitly shown in that figure, one can interpret that tree as representing the antecedent (*if* part) of an implicit classification rule, which will assign, to examples satisfying the Boolean function represented in that tree, a certain class corresponding to the implicit consequent (*then* part) of the rule. In addition, although most GP algorithms seem to focus on discovering just one classification rule for each class (which might be a very complex rule, due to the need to cover all examples of a class), it is possible to use a sequential covering approach (as discussed in the context of conventional rule induction algorithms in

Table 4.1 Summary of function and terminal sets of several GP algorithms for evolving classification models in the form of a classification function or classification rules

reference	main types of terminal set elements	main types of function set elements
[3]	real-valued attributes	mathematical operators
[4]	real-valued attributes	mathematical operators
[13]	real-valued attributes	mathematical operators, if
[22]	real-valued attributes	mathematical operators
[25]	real-valued attributes	mathematical operators, *soft* if
[37]	real-valued attributes	mathematical operators, *soft* if
[27]	real-valued attributes	fuzzy logic operators
[9]	booleanized attributes	logical operators
[14]	booleanized attributes	logical operators
[21]	alphanumeric characters	string-handling and logical operators

Chapter 2) to discover several rules for each class, where each rule could then be simpler since it would have to cover just a subset of examples of the target class.

Table 4.1 presents a summary of several works evolving classification models in the form of a conventional GP tree that can be interpreted as a classification function or a classification rule. For each work, this table shows the main types of elements used in the terminal and function sets of the GP. In the last column (function set elements), the term "*soft* if" refers to a modified type of *if* statement that returns a numerical value, instead of a Boolean value as usual, in order to satisfy the closure property in the context of other operators that accept only a numerical value as an input.

4.3.2 Evolving Decision Trees

Let us now turn to GP algorithms for evolving decision trees. A decision tree is of course a type of classification model, but this type of GP algorithm is significantly different from the standard type of GP algorithm for evolving a classification model discussed in the previous subsection. The main difference is that a decision tree has a specific type of structure, where each leaf node contains a class to be predicted for a new example and each internal node is associated with a test based on the values of one or more attributes. One can view the attribute value tests associated with internal nodes as functions, and so the function set (in GP terminology) can be thought of as including the set of predictor attributes of the data being mined and some comparison operators (e.g. $=$, $>$) used to implement the attribute value tests. In addition, the terminal set (in GP terminology) consists of the classes to be predicted. A simple example of a decision tree (which could be represented by a GP individual or have been built by a more conventional decision tree induction method, as discussed in Section 2.3) is shown in Fig. 4.2(b).

Note that this composition of the function and terminal sets in a GP algorithm for evolving a decision tree is very different from its counterpart in a conventional

GP tree used as a classification function. In the latter, as discussed in the previous subsection, the predictor attributes are in the leaf nodes, and the classes typically do not appear in the tree representing an individual at all; the class to be assigned to a new example will depend on the value returned by the individual's decision tree's root node. These differences are illustrated in Fig. 4.2.

Another difference is that, when the classification model is expressed as a decision tree, the classification of a new example is done in a top-down fashion, first showing the example to the root node and then passing the example downwards the tree until a leaf node is reached and the corresponding class prediction is made. In contrast, when the classification model is expressed as a conventional GP tree, the classification of a new example is done in a bottom-up fashion by instantiating the individual's leaf nodes' attributes with their corresponding values in the example and propagating that information upwards in the tree until a value is output at the root node and converted into a predicted class, as explained earlier.

The broad group of GP algorithms for evolving decision trees can be further subdivided into two subgroups according to whether the decision tree being evolved contains only univariate tests in their internal nodes (the most popular type of decision tree) or whether the tree being evolved contains multivariate tests in its internal nodes.

Recall that, in a conventional decision tree, each internal node is associated with with a test based on the values of a single attribute (univariate test). Most GP algorithms designed for evolving decision trees produce decision trees containing only such univariate tests; see, e.g., [6, 16, 19, 26, 31, 36].

However, it is also possible to have multivariate tests in a decision tree's internal nodes. Multivariate tests tend to be used less often because they are more complex, producing trees that are less understandable to the user and increasing the computational time required to search for the best attribute test at each node. On the other hand, a multivariate test has a considerably greater expressiveness power and tends to cope better with attribute interaction than a univariate test. GP algorithms that evolve decision trees with multivariate tests are described in [10, 12, 35].

4.4 Genetic Programming for Evolving Components of Rule Induction Algorithms

Recall that a rule induction algorithm essentially consists of several types of components, in particular the following major components (the list below ignores other algorithm components irrelevant for our discussion, such as the rule pruning component):

- the representation of the candidate rules;
- the search strategy used to explore the space of candidate rules;
- the evaluation function used to measure the quality of candidate rules.

Wong [41] proposed a grammar-based GP system for evolving the evaluation function of a rule induction algorithm. Hence, that GP system can be said to automate the partial design of a classification algorithm, since just one component of the algorithm was automatically evolved by the GP system.

The rule induction algorithm used in [41] was an Inductive Logic Programming (ILP) algorithm (based on the FOIL algorithm [32]). Most components of the ILP algorithm were fixed during the experiments rather than being evolved by the GP system. Those fixed components were specified as follows. The representation was first-order logic, which is the major characteristic of ILP algorithms. This means the discovered rules can have first-order logic or relational rule conditions such as *Income* > *Expenses*, involving a relational comparison between two attributes. This is in contrast to the conventional propositional logic representation, where each rule condition can only refer to a single attribute and values belonging to the domain of that attribute, such as *Income* > *50,000*. The search strategy used to explore the space of candidate rules involved a top-down rule refinement approach, which starts from general candidate rules and iteratively transforms them into more specific rules using rule specialization operators, and a sequential covering approach. Both approaches are popular in rule induction algorithms.

Let us now turn to the single component of the ILP algorithm that was not fixed, but rather was evolved by the GP system in [41]. The component in question was the rule evaluation function, i.e., a function that accepts as input a candidate classification rule and produces as output a numerical value indicating the quality of that rule. Rule quality is typically measured in terms of the classification accuracy of the rule, but there is no consensus about precisely which formula should be used to measure rule quality (see Section 2.4.3). A major problem is that, although our ultimate goal is to discover rules with the highest possible predictive accuracy (generalization ability) in the test set, unseen during training, we cannot, by definition of the classification task, use the test examples to compute the rule quality. That is, rule quality has to be estimated by using just training examples, and there are many different formulas proposed in the literature to compute such an estimate.

Actually, each rule evaluation formula has a certain evaluation bias, which can be considered a kind of "search bias" [40] in the sense that it influences the search performed by the rule induction algorithm. Since in general the effectiveness of a search bias is dependent on the dataset being mined, it follows that the effectiveness of a given rule evaluation formula is dependent on the dataset. Instead of trying to find out which of the many formulas proposed in the literature is more suitable to the dataset being mined, Wong [41] proposed a GP system to evolve a new rule evaluation function tailored to the dataset being mined in a data-driven way. That is, the system performs rule evaluation function *construction*, rather than rule evaluation function *selection*.

For the purposes of our discussion here, the most important elements of the proposed GP system are its terminal set and its function set. Let "positive class" be the class predicted by a given candidate rule and "negative class" be any class different from the class predicted by that rule. The terminal set involves elements such as

- n^+: number of positive-class training examples that satisfy the antecedent of the current candidate rule;
- n^-: number of negative-class training examples that satisfy the antecedent of the current candidate rule;
- a random number generator that generates a number between -10 and 10;
- the amount of information (based on information theory) of the class distribution associated with the current candidate rule: $info = -log_2(n^+/(n^+ + n^-))$.

Note that, although the above third and fourth elements look like functions at first glance, they are actually terminal symbols from the GP system's point of view, since they are represented by leaf nodes in an individual's tree (which represents a candidate rule evaluation function). Intuitively a good rule evaluation function should assign better scores to rules with larger values of n^+ and smaller values of n^-, but these and other elements of the terminal set can be combined in many different ways by functions and operators of the function set in order to create new rule evaluation functions.

The function set of the proposed GP system involves arithmetic operators such as "+", "−", "×", "protected division" (which returns the quotient of the division as usual, except in the case of division by zero, when it returns 1), and "protected logarithm" (which returns the log of the input argument x if x is greater than zero, and returns 1 otherwise). For details of the function and terminal sets and how they are combined in a grammar for generating candidate rule evaluation functions, see [41].

The important point about this work in the context of this book is that the GP system is used to automate part of the design of a classification *algorithm*, rather than being used to evolve a classification *model*. Recall the crucial difference between a classification algorithm and a classification model explained in Section 4.2. Let us elaborate on this point.

First, note that in general (with the exception of the random number generator) the elements of the terminal set refer to variables representing abstract performance statistics of a candidate rule, i.e., variables related to the number of positive and negative examples satisfying the antecedent of a candidate rule. It is important to note that the definition of this type of variable is completely independent of the particular application domain or dataset being mined. That is, these variables have the same meaning in any classification dataset, regardless of the type of application domain (medicine, finance, bioinformatics, or any other) of the data being mined. Hence, the evolved rule evaluation function is indeed a component of a truly generic ILP algorithm, since the ILP algorithm with the evolved rule evaluation can be applied to many different types of classification datasets belonging to many different types of application domains, in the same way that a completely human-designed ILP algorithm can be applied to many different types of classification datasets and application domains.

This should be contrasted with the GP algorithms for evolving the classification model mentioned in Section 4.3, where the terminal and function sets refer to dataset-specific variables. In particular, in the case of GP algorithms evolving a classification model in the form of a classification function (Section 4.3.1) the terminal set includes the names of predictor attributes of the data being mined, while in the case of GP algorithms for evolving a classification model in the form of a decision tree (Section 4.3.2) the terminal set contains the names of the classes and the function set contains the names of predictor attributes of the data being mined. The use of dataset-specific variables is a natural consequence of the fact that in the GP algorithms mentioned in Section 4.3 (as in almost any work using GP in data mining) what is being evolved is a classification model specific to the application domain whose dataset is being mined. For instance, it does not make any sense at all to apply a classification *model* induced from data about a medical application (with attributes describing a patient's characteristics) to data about a financial application (with attributes describing a customer's characteristics). In contrast, as mentioned earlier, the same classification *algorithm* can easily be (and very often is) applied to datasets from very different application domains.

At this point in the discussion, it is important not to get confused with two different types of generality of a rule evaluation function. As we just said, in the work discussed in this section the GP system evolves a rule evaluation function that is generic in the sense that it can be applied to datasets from different application domains. We also said earlier that the evolved rule evaluation function is tailored to the dataset being mined. These two statements are *not* inconsistent; they simply refer to different types of generality. The former statement involves generality with respect to the "legal" applicability of the evolved funtion, i.e., from a syntactical and semantical point of view the evolved rule evaluation function can be applied to classification datasets from different application domains. The latter statement refers to the expected specificity of the evolved rule evaluation function from a performance point of view. That is, the evolved function is tailored to the dataset being mined, and so it is expected to be a better estimate of predictive accuracy in the specific dataset being mined than other rule evaluation functions that were not tailored to that dataset.

To summarize, Wong's proposed GP system automatically designs a rule evaluation function for a rule induction algorithm, where the evolved function is so generic that it can be applied to evaluate rules in classification datasets from different application domains. However, Wong's work still has the limitation that just one component of a rule induction algorithm (its rule evaluation function) is evolved by the GP system, characterizing the automation of the partial design of a rule induction algorithm. In contrast, the GP system proposed in this book automatically designs a complete rule induction algorithm, as will be discussed in detail in Chapter 5.

4.5 Genetic Programming for Evolving Classification Systems

Suyama et al. [38] proposed a hybrid of a GP algorithm and a local search method to evolve a classification system. Their proposed system, called CAMLET, differs from a conventional GP algorithm in the sense that it does not use traditional GP operators, but rather three strategies that use heuristics to generate new individuals. These strategies were named greedy alteration, random generation, and heuristic alteration. The first two strategies are adaptations of crossover and mutation operators, while the last one is a kind of local search. In a more recent work, Abe and Yamaguchi [1] compared a parallel version of CAMLET to three stacking methods, but replaced its hybrid GP and local search mechanism by a simple genetic algorithm, using tournament selection, crossover, and mutation.

It is important to note that CAMLET aims at automatically designing a classification system (which the authors referred to as an "inductive learning system") in a broad sense, because it designs both dataset-related and algorithm-related aspects of a classification system. In order to guide the design of both these aspects of a classification system, CAMLET uses an ontology rather than a grammar. The ontology used in [38] includes 15 coarse-grained building blocks at the terminal (leaf) nodes of the ontology. These building blocks are coarse-grained in the sense that in general they correspond to complete (or almost complete) classification algorithms or to coarse-grained decisions about dataset design.

For instance, dataset-related aspects of the classification system being evolved include issues such as how to generate training and testing sets from the available dataset, and corresponding building blocks in terminal nodes of the ontology represent coarse-grained decisions such as whether the process of dataset generation uses bootstrapping (allowing duplication of examples) or not.

Concerning the issue of classification algorithm design in CAMLET, this system can construct several types of classification algorithms (e.g., rule induction, decision trees, neural networks), facilitated by the coarse-grained level of its building blocks. The coarse-grained nature of the ontology is particularly noticeable in the case of the building blocks corresponding to classification algorithms, since in general each type of classification algorithm is implemented either as a single terminal node in the ontology or as a couple of terminal nodes corresponding to different versions of that type of algorithm – where the versions differ with respect to some coarse-grained issue in the design of the algorithm. For instance, based on the brief description of the ontology given in [38, 1], it seems that the system can produce only two versions of a decision tree algorithm (represented by two terminal nodes in the ontology) differing in their attribute evaluation functions; and it seems that a single version of a rule induction algorithm (represented by a single terminal node) can be produced.

To summarize, CAMLET uses GP to automatically design a full classification system, but with three basic differences from the GP system proposed in this book to automatically design rule induction algorithms – to be described in detail in Chapter 5. First, CAMLET aims at designing both dataset-related and algorithm-related aspects of a classification system, while the GP system proposed in this book aims at designing just a classification algorithm, and not dataset-related aspects of a broader system. Secondly, in order to construct new classification systems, CAMLET manipulates building blocks that are defined at a coarse-grained level of abstraction; in some cases a full or almost full classification algorithm seems to be a "building block" for the broader classification system. In contrast, the proposed GP system uses much more fine-grained building blocks, including programming constructs like "while," "if,", etc., as well as several fine-grained rule induction procedures (not available in CAMLET), as will be seen later. Thirdly, CAMLET can construct several types of classification algorithms (e.g., rule induction, decision trees, neural networks), facilitated by the coarse-grained level of its building blocks. In contrast, the proposed GP system constructs only rule induction algorithms, since it has a finer-grained grammar specialized for the creation of rule induction algorithms.

4.6 Evolving the Design of Optimization Algorithms

The focus of this book is on data mining, in particular on the classification task. Hence, the reader may wonder why we included in this chapter a section on optimization (focusing on combinatorial optimization), since this type of problem is quite different from classification. The motivation for this section is twofold. First, although classification and optimization are different problems (as explained in more detail below), both involve a search in the space of candidate solutions, and at a high level of abstraction some concepts and principles in one of these fields may be transferred to the other field. Secondly, since there is little work done on automating the design of data mining algorithms – which includes the GP system described in the next chapter and the works discussed in Sections 4.4 and 4.5 – we believe it makes sense to expand this chapter's discussion to consider methods for automating the design of optimization algorithms. This is because the latter is a research area that, overall, seems somewhat more developed (having more published work) than the area of automating the design of classification algorithms. Before we move on to discuss methods for automating the design of optimization algorithms, though, we first review important differences between optimization and classification.

4.6.1 Optimization Versus Classification

The first difference between optimization and classification is the fact that optimization is usually a "well-defined," deterministic task, while classification is almost by definition an "ill-defined," nondeterministic task [17], in the following sense. In most optimization problems, given a problem instance, there is a well-defined (set of) optimal solutions for that problem. Consider, for instance, the well-known Traveling Salesman Problem, where one has to find the shortest tour passing exactly once through each city. Once both the set of cities and the distances between each pair of cities has been specified, this specification precisely determines the optimal solutions – shortest tours – to be found by an optimization method. (Let us ignore, for the purposes of our discussion here, less common nondeterministic versions of the Traveling Salesman Problem, such as "blind," or "noisy," versions of the problem where the distance between some pairs of cities is unknown before a tour involving those cities is completed.)

In contrast, in the classification task, the system is given a finite amount of training data, and it has to use that data to induce a model that hopefully will make accurate class predictions in the test data, unseen during training. Hence, even when we use a deterministic classification algorithm, the actual classification model induced from a training set will depend on which sample of examples was included in the training set. In addition, the predictive accuracy measured by applying the induced classification model to the test examples will also depend on which sample of examples was included in the test set. (Both the training set and the test set contain just a finite sample of examples drawn from a potentially infinite example space.) These points characterize a nondeterministic task.

The second difference between optimization and classification is that classification systems must find a solution with a good generalization ability (predictive performance on a test set unseen during training), while normally there is no requirement for generalization ability in the case of conventional optimization systems. Note that the situation is different in the context of evolutionary algorithms for automating the design of optimization *algorithms*, as will be discussed in Section 4.6.3, where the evolved optimization algorithm should have a good generalization ability. However, in this case finding a solution with good generalization ability is a goal for the evolutionary algorithm, which performs its search in the space of optimization algorithms, rather than a goal for the optimization algorithm, which simply has to find the optimal solutions for a given instance of the target problem, without concerns about generalization ability.

An important point related to generalization ability is that classification systems have to try to avoid overfitting (and its complement, underfitting), which is not an issue, in general, in optimization problems. Recall from Section 2.2.2 that the term overfitting refers to the case where a classification model is adjusted to details of the training data so much that it is representing very specific idiosyncracies of that data, and so it is unlikely to generalize well to the test data (unseen during training). In conventional optimization, overfitting is not an issue because there is no separation between "training" and "testing" phases, and so there is no goal of discovering a model that generalizes beyond the current data.

The third difference is that oversearching is an important issue in classification, while it is not an issue in optimization. Actually, oversearching is one possible cause of the previously mentioned problem of overfitting, but the former is important enough in the context of evolutionary algorithms in order to be discussed separately. In essence, oversearching occurs when the classification algorithm evaluates too many candidate classification models, so that the best model found by the algorithm (according to the criterion of consistency with the training set) represents a spurious (or "fluke") relationship that is unlikely to generalize well for the test set, unseen during training. Again, oversearching is not an issue in optimization because in that type of problem there is no separation between "training" and "testing" phases, and no need for generalization. Let us elaborate on this point.

Suppose we are given an instance of a well-defined, deterministic combinatorial optimization problem such as the Traveling Salesman Problem, and we use an

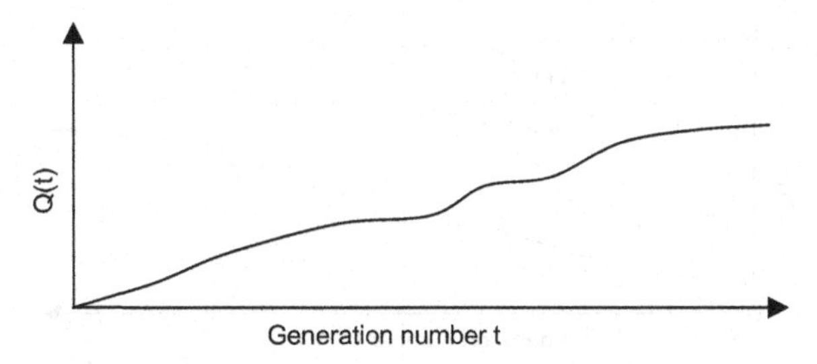

Fig. 4.3 Quality of the best solution found so far across generations in optimization (where oversearching is not an issue)

evolutionary algorithm to solve that problem instance. Consider an evolutionary algorithm using the simple technique of "elitism," where, at every generation, the best individual of that generation, i.e., the individual with the best fitness value, is selected and copied to the next generation without any modification, i.e., without undergoing crossover, mutation, or any other genetic operation. It is clear that, in this case, the quality (fitness) of the best solution found by the algorithm so far will be monotonically non-decreasing along the generations. That is, for every generation (time index) t, $Q(t+1) \geq Q(t)$, where $Q(t)$ denotes the quality (fitness) of the best individual at generation t and it is assumed, without loss of generality, that the function $Q(t)$ is to be maximized. In this scenario, if we increased significantly the number of generations of the evolutionary algorithm, this would make the algorithm take significantly more processing time, but this would not decrease the quality of the solution found by the algorithm. This is illustrated in Fig. 4.3.

The situation is very different in the case of classification, due to the need for discovering a classification model that generalizes as well as possible to the test set. In this case, if we use an evolutionary algorithm with elitism as described above, it is still true that for every generation t, $Q(t+1) \geq Q(t)$. However, it is important to note that in this case the quality (fitness) function Q would be measuring the classification accuracy of a given individual (candidate classification model) on the training set. This is, by definition, just an imperfect approximation of the quality measure that we really want to maximize, which is the predictive accuracy on the test set, but which cannot be directly maximized by the evolutionary algorithm, since it cannot access the test set. Hence, when using an elitist evolutionary algorithm for classification, although the fitness of the best individual in the current generation will never decrease when the number of generations is increased, the predictive accuracy on the test set of the best individual in the current generation may well decrease after a certain number of generations due to oversearching. This is illustrated in Fig. 4.4, where t_{opt} would be the optimal number of generations (unknown by the evolutionary algorithm) at which we would ideally like to stop the run of the evolutionary algorithm in order to maximize predictive accuracy in the test set.

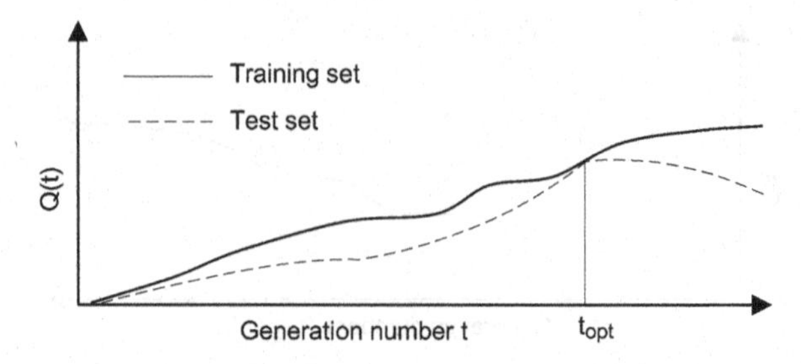

Fig. 4.4 Quality of the best solution found so far across generations in classification (where over-searching is an issue)

4.6.2 On Meta-heuristics and Hyper-heuristics

Research on automating the design of optimization algorithms has been done mainly using the so-called "hyper-heuristics," which are a special kind of "meta-heuristics." To understand the meaning of these terms, let us first briefly discuss meta-heuristics, starting with a discussion of the more basic term "heuristic."

Broadly speaking, the term heuristic is normally used to refer to a "rule of thumb" or an "educated guess" to solve a certain type of problem. (The name "heuristic" comes from a Greek word, *heuriskein*, which means "to find.") Typically, although a heuristic does not guarantee the discovery of an optimal solution, a heuristic should at least allow us to discover a good (perhaps satisfactory) solution in a period of time much shorter than the time that would be taken by a method that guarantees the discovery of the optimal solution; otherwise, there would not be much motivation to use a heuristic, since we could simply use a method guaranteeing the optimal solution instead. For instance, in the Traveling Salesman Problem, a simple heuristic to construct a tour consists of choosing, as the next city to be visited, the unvisited city which is nearest to the last city that was added to the current partially constructed tour.

The term meta-heuristics was coined to denote a "meta-level" (higher-level) heuristic. Several definitions of the term meta-heuristic are given in [7], out of which we will quote the following one: "A metaheuristic is a set of concepts that can be used to define heuristic methods that can be applied to a wide set of different problems. In other words, a metaheuristic can be seen as a general algorithmic framework which can be applied to different optimization problems with relatively few modifications to make them adapted to a specific problem" [7] (p. 270).

Evolutionary algorithms are considered a type of meta-heuristic in the optimization literature, as are other population-based search methods such as ant colony optimization and particle swarm optimization.

A hyper-heuristic can be broadly defined as "the process of using (meta-) heuristics to choose (meta-) heuristics to solve the problem in hand" [11].

An important difference between meta-heuristics and hyper-heuristics is that meta-heuristics perform a search in the space of candidate solutions for the target problem, while hyper-heuristics perform a search in the space of heuristics that can be used to solve a given type of problem. Most current hyper-heuristics are based on evolutionary algorithms (modified to do a search in the space of heuristics), although in principle any other type of meta-heuristic could be properly modified and used as a hyper-heuristic too.

At this point a comment about terminology is relevant. The terms "meta-heuristic" and "hyper-heuristic" seem inappropriate and/or confusing in a general computer science context. In this context, the term "meta" is usually used to refer to a higher-level process or object of the same fundamental nature as its corresponding counterpart at the base level. For instance, in the area of database systems, meta-data is essentially data about data. Hence, a heuristic that is used to choose another heuristic should be called a "meta-heuristic," rather than a "hyper-heuristic." In addition, what is normally called a "meta-heuristic" does not seem to deserve that name, and maybe a better name would be "generic heuristic," or perhaps "problem-independent heuristic." After all, in general current meta-heuristics do not explicitly operate on or control lower-level heuristics; they are simply a kind of heuristic that have a much broader applicability than problem-specific heuristics. Having said that, the fact is that the term meta-heuristic is already deeply embedded in the optimization literature, and the term hyper-heuristic is starting to become popular too. Therefore, for the sake of consistency with the literature, we use these terms as they are currently used in the optimization literature throughout this section.

4.6.3 Evolving the Core Heuristic of Optimization Algorithms

We discuss here hyper-heuristics for evolving the core heuristic component of optimization algorithms for two types of combinatorial optimization problems, the Traveling Salesman Problem and the Satisfiability problem.

4.6.3.1 Evolving a Heuristic for the Traveling Salesman Problem

Oltean and Dumitrescu [30] proposed a type of GP algorithm (specifically, multi-expression programming) for evolving a heuristic function for the Traveling Salesman Problem. More precisely, the proposed algorithm evolves the core component of an algorithm for solving the Traveling Salesman Problem, which is a heuristic function that evaluates the candidate cities to be added to the current partially constructed tour. During the construction of a tour, this type of function is used to decide which city, out of all cities unvisited so far, should be added to that partially constructed tour. Since what is evolved is an evaluation function (a mathematical expression) rather than a full algorithm or program, the function set consists essentially of mathematical operators. Hence, there is no need to have programming constructs such as loop or conditional statements in the function set.

In any case, the GP algorithm does discover solutions that are expressed at a considerably higher level of generality than the solutions discovered by conventional evolutionary algorithms for the Traveling Salesman Problem. This can be explained as follows. Conventional evolutionary algorithms for the Traveling Salesman Problem return, as a solution, the best found tour for a single instance of the Traveling Salesman Problem, defined by a single predefined set of cities with predefined weighted edges between them. In contrast, the GP algorithm proposed by Oltean and Dumitrescu returns, as a solution, the best found evaluation function for the Traveling Salesman Problem in general, a solution that is applicable to potentially any instance of the Traveling Salesman Problem, i.e., any set of cities with any set of weighted edges between them. Actually, the fitness function of the GP algorithm was computed by applying the candidate heuristic function represented by an individual to several randomly chosen instances of the Traveling Salesman Problem, in order to try to evolve evaluation functions that are robust to different instances of that problem.

4.6.3.2 Evolving a Heuristic for the Satisfiability (SAT) Problem

Fukunaga [20] proposed a kind of GP algorithm for evolving heuristics for the SAT (satisfiability) problem, a well-known type of combinatorial optimization problem. The basic definition of a SAT problem is as follows. The system is given, as input, a Boolean expression, i.e., a logical expression where each variable is Boolean, taking only a true or false value. The basic goal of the system is to determine whether or not it is possible to assign values to the variables in the given Boolean expression in a way that makes that expression true. If such an assignment of values exists, the expression is said to be satisfiable, and in this case the system should return the sets of variable assignments that made the Boolean expression true.

Fukunaga's work consists of evolving the core component of a well-known type of local search method for the SAT problem, which starts with a certain assignment of values to the variables of the given Boolean expression and then iteratively flips one variable at a time until either the Boolean expression is made satisfiable or a predefined number of flips (the maximum number allowed) is exceeded. The initial assignment of values to the variables is typically randomly generated, but the choice of which variable should be flipped at each iteration is determined by a heuristic, called a variable-selection heuristic, which is the core component of this type of local search method. In SAT research, the variable-selection heuristic is typically manually designed, and the important contribution of Fukunaga's work was to propose an automated method for discovering such heuristics.

The proposed automated method is based on two key ideas. First, several structural elements that were common among a number of known (manually designed) variable-selection heuristics were identified. Secondly, a kind of GP algorithm was proposed to automatically combine those common structural elements ("building blocks") into a new variable-selection heuristic for the SAT problem. In the proposed algorithm the individuals in the initial population are created using a grammar, which enforces the constraint that each GP individual is a syntactically valid heuristic that returns the index of the selected variable when executed. The main operator used by that GP algorithm is a composition operator, which essentially combines two variable-selection heuristics H_1 and H_2 into a new composite (meta-level) heuristic that chooses whether H_1 or H_2 should be executed based on a certain condition. Hence, this operator has the form: *if* (*Condition*) *then* (H_1) *else* (H_2). The *Condition* element of the composition operator can take several forms, but basically it refers to a scoring of the Boolean variables based on the desirability of choosing each variable as the next variable to be flipped in the current iteration of the local search method.

In order to compute the quality of a given candidate heuristic, a SAT problem-solving algorithm using that heuristic is run on a number of randomly generated instances of the SAT problem. The quality of a candidate heuristic is computed as a function of two factors, namely (a) how many of the SAT problem instances were correctly solved, and (b) the mean number of flips of variables that were performed in order to solve the SAT problem instances.

Note that in his paper Fukunaga did not present his proposed algorithm as a GP algorithm; he rather referred to it as a population-based search algorithm. However, after presenting the algorithm, he mentioned that it can be viewed as a grammar-based GP system using only a novel type of composition operator as the search operator, rather than the standard genetic operators used in conventional GP algorithms.

A more conventional type of grammar-based GP system for evolving a variable-selection heuristic for the SAT problem was proposed by Bader-El-Den and Poli [5]. One of the motivations for proposing this new system, as pointed out by the authors, is that in Fukunaga's work the use of conditional operators as the only type of search operators "did not give the ... system enough freedom to evolve heuristics radically different from the human-designed heuristics (effectively, the evolved heuristics are made up by a number of nested heuristics)" [5, p. 40].

In order to evolve heuristics that were more different from human-designed ones than the heuristics evolved by Fukunaga's GP system, Bader-El-Den and Poli proposed what they considered a flexible grammar, which seems to allow the generation of a wider set of heuristics than does the grammar used by Fukunaga. In their GP system, more conventional grammar-based crossover and mutation operators are used, so that the genetic operators always produce individuals representing syntactically valid heuristics.

In order to compute the fitness of a GP individual, a SAT problem-solving algorithm with the heuristic represented by that individual is run on a number of instances of the SAT problem. The individual's fitness is then computed as a function of three factors, namely (a) how many of the SAT problem instances were correctly solved, (b) the mean number of flips of variables that were performed in order to solve the SAT problem instances, and (c) how many nodes are contained in the individual's tree representing a candidate heuristic. The larger the value of factor (a), and the smaller the values of factors (b) and (c), the better the fitness of an individual.

It should be noted that the development of a GP system for automating the design of classification algorithms could borrow some ideas from the fitness functions used in [20] and [5], as follows. As will be discussed in Chapter 5, the GP system for automating the design of rule induction algorithms proposed in that chapter uses, in its fitness function, evaluation criteria related to the quality of the solution (classification model) produced by an automatically designed rule induction algorithm, but not criteria related to the efficiency (processing time) or simplicity of that rule induction algorithm. In contrast, the GP systems proposed in [20] and [5] use a fitness function taking into account not only the quality of the solution produced by the automatically designed SAT-solving algorithm (how many of the SAT problem instances were correctly solved), but also criteria related to the efficiency of the SAT-solving algorithm (the mean number of flips of variables performed) and – in the case of Bader-El-Den and Poli's work – that algorithm's "simplicity" (number of nodes contained in the individual's tree representing the algorithm). Using these latter two types of algorithm evaluation criteria in the fitness function of the GP system proposed in Chapter 5 would be an interesting research direction.

4.6.4 Evolving an Evolutionary Algorithm for Optimization

Oltean [29] proposed a (meta-level) evolutionary algorithm for designing a (base-level) evolutionary algorithm that solves a given target problem. That is, the former is a very generic evolutionary algorithm working at a meta-level (or macro-level in the author's terminology), and the solution returned by that meta-level evolutionary algorithm is an evolutionary algorithm working at the level of the problem being solved (called the micro-level in the author's terminology). In [29] the meta-level evolutionary algorithm is used to evolve base-level evolutionary algorithms for three different types of problems, namely a numerical function optimization problem (where the goal is to find the optimal value of a predefined function) and two well-known combinatorial optimization problems, viz. the Traveling Salesman Problem and the Quadratic Assignment Problem. Note that each run of the meta-level evolutionary algorithm evolves a base-level evolutionary algorithm for just one of these three types of problems, rather than considering all the three problems at the same time.

The basic idea of the meta-level evolutionary algorithm is that its function set consists of the genetic operators that will be used by the base-level evolutionary algorithm. In other words, in the meta-level evolutionary algorithm an individual will specify which kind of genetic operators such as crossover and mutation will be implemented in its corresponding candidate base-level evolutionary algorithm. The precise definition of which kinds of crossover and mutation operators will compose the function set is to some extent dependent on the problem that has to be solved by the base-level evolutionary algorithm. For instance, the Traveling Salesman Problem usually requires specialized types of crossover capable of handling candidate solutions in the form of permutations [15, 39]. In any case, there are several types of crossover that can be applied to the Traveling Salesman Problem, and instead of manually choosing which one should be used, that decision (with other decisions about the design of the base-level evolutionary algorithm) is left to be automatically evolved by the meta-level evolutionary algorithm. The meta-level evolutionary algorithm can also be used to evolve parameters of the base-level evolutionary algorithm, such as the latter's population size, number of generations, and probability of crossover and mutation.

The fitness of an individual in the meta-level evolutionary algorithm is computed by running the base-level evolutionary algorithm represented by that individual on the target problem. Since evolutionary algorithms are nondeterministic, the base-level evolutionary algorithm is run a number of times on the target problem, and the average fitness value of the solutions returned by those runs is used as the fitness of the corresponding individual of the meta-level algorithm.

4.7 Summary

This chapter started with a discussion of the differences between classification models (the output of classification algorithms) and classification algorithms themselves. This is a crucial point to understand the contribution of this book, because the vast majority of the GP literature addresses the problem of evolving classification models, while in this book we address the more difficult problem of evolving a classification algorithm. We then presented a brief review on the use of GP for evolving classification models represented in the form of classification functions (possibly interpreted as classification rules) or decision trees, since these are the types of classification models more used in the GP literature.

Next we discussed two kinds of related work on using GP for automatically designing (aspects of) classification algorithms. First, we discussed Wong's work [41], proposing a GP system to automatically design a rule evaluation function for a rule induction algorithm. The main difference between that system and the GP system proposed in this book – to be discussed in detail in Chapter 5 – is that the latter automatically designs a complete rule induction algorithm rather than just its evaluation function. Secondly, we discussed the CAMLET system proposed by Suyama et al. [38] and further investigated by Abe and Yamaguchi [1]. This system uses GP to automatically design a full classification system – involving the design of both dataset-related and algorithm-related aspects of a classification system. However, as discussed earlier, there are several basic differences between CAMLET and the GP system proposed in this book. The most important of these differences seems to be that CAMLET manipulates building blocks that are defined at a coarser-grained level of abstraction, in comparison with the finer-grained building blocks of the grammar used by the proposed GP system.

Finally, since there is so little research on automating the design of classification algorithms, we included in this chapter an overview of the research on using GP to automatically design (aspects of) optimization algorithms (mainly combinatorial optimization algorithms), a research area that is currently more developed than the area of automating the design of classification algorithms.

References

1. Abe, H., Yamaguchi, T.: Comparing the parallel automatic composition of inductive applications with stacking methods. In: R. Camacho, A. Srinivasan (eds.) Proc. of the Workshop on Parallel and Distributed Computing for Machine Learning (ECML/PKDD-03), pp. 1–12. Cavtat-Dubrovnik, Croatia (2003)
2. Abraham, A.: Meta learning evolutionary artificial neural networks. Neurocomputing **56**, 1–38 (2004)
3. Almal, A., Mitra, A., Datar, R., Lenehan, P., Fry, D., Cote, R., Worzel, W.: Using genetic programming to classify node positive patients in bladder cancer. In: Proc. of Genetic and Evolutionary Computation Conf. (GECCO-06), pp. 239–246. Morgan Kaufmann (2006)
4. Archetti, F., Lanzani, S., Messina, E., Vanneschi, L.: Genetic programming for human oral bioavailability of drugs. In: Proc. of Genetic and Evolutionary Computation Conf. (GECCO-06), pp. 255–262. Morgan Kaufmann (2006)
5. Bader-El-Den, M., Poli, R.: Generating SAT local-search heuristics using a GP hyper-heuristic framework. In: Artificial Evolution (Proc. of 8th Int. Conf. on Evolution Artificielle), *LNCS*, vol. 4926, pp. 37–49. Springer-Verlag (2007)
6. Basgalupp, M., Barros, R., Carvalho, A., Freitas, A., Ruiz, D.: Legal-tree: a lexicographic multi-objective genetic algorithm for decision tree induction. In: Proc. of 24th Annual ACM Symposium on Applied Computing (SAC 2009), Hawaii, USA, pp. 1085–1090 (2009)
7. Blum, C., Roli, A.: Metaheuristics in combinatorial optimization: overview and conceptual comparison. ACM Computing Surveys **35**(3), 268–308 (2003)
8. Bojarczuk, C., Lopes, H., Freitas, A.: Discovering comprehensible classification rules using genetic programming: a case study in a medical domain. In: Proc. of the Genetic and Evolutionary Computation Conf. (GECCO-99), pp. 953–958. Morgan Kaufmann (1999)
9. Bojarczuk, C.C., Lopes, H.S., Freitas, A.A.: Genetic programming for knowledge discovery in chest pain diagnosis. IEEE Engineering in Medicine and Biology Magazine **19**(4), 38–44 (2000)
10. Bot, M., Langdon, W.: Application of genetic programming to induction of linear classification trees. In: Proc. of the 3rd European Conf. on Genetic Programming (EuroGP-00), *LNCS*, vol. 1802, pp. 247–258. Springer (2000)
11. Burke, E., Kendall, G., Newall, J., Hart, E., Ross, P., Schulenburg, S.: Hyper-heuristics: an emerging direction in modern search technology. In: F. Glover, G. Kochenberger (eds.) Handbook of Meta-Heuristics, pp. 457–474. Kluwer (2003)
12. Cantu-Paz, E., Kamath, C.: Using evolutionary algorithms to induce oblique decision trees. In: Proc. of the Genetic and Evolutionary Computation Conf. (GECCO-00), pp. 1053–1060. Morgan Kaufmann (2000)
13. Cavaretta, M.J., Chellapilla, K.: Data mining using genetic programming: the implications of parsimony on generalization error. In: P.J. Angeline, Z. Michalewicz, M. Schoenauer, X. Yao, A. Zalzala (eds.) Proc. of the Congress on Evolutionary Computation (CEC-99), vol. 2, pp. 1330–1337. IEEE Press (1999)
14. Eggermont, J., Eiben, A., van Hemert, J.: A comparison of genetic programming variants for data classification. In: Proc. of Conf. on Intelligent Data Analysis (EuroGP-99). Springer (1999)
15. Eiben, A.E., Smith, J.E.: Introduction to Evolutionary Computation. Springer-Verlag (2003)
16. Folino, G., Pizzyti, C., Spezzano, G.: Genetic programming and simulated annealing: a hybrid method to evolve decision trees. In: Proc. of the 3rd European Conf. on Genetic Programming (EuroGP-00), *LNCS*, vol. 1802, pp. 294–303. Springer (2000)

17. Freitas, A.A.: Understanding the crucial differences between classification and discovery of association rules: a position paper. ACM SIGKDD Explorations **2**(1), 65–69 (2000)
18. Freitas, A.A.: Data Mining and Knowledge Discovery with Evolutionary Algorithms. Springer-Verlag (2002)
19. Fu, Z.: An innovative ga-based decision tree classifier in large scale data mining. In: Proc. of the 3rd European Conf. on Principles and Practice of Knowledge Discovery in Databases (PKDD-99), pp. 348–353. Springer (1999)
20. Fukunaga, A.: Automated discovery of composite SAT variable-selection heuristics. In: Proc. of the National Conf. on Artificial Intelligence (AAAI-02), pp. 641–648. AAAI Press (2002)
21. Hirsch, L., Saeedi, M., Hirsch, R.: Evolving text classifiers with genetic programming. In: Proc. of the 7th European Conf. on Genetic Programming (EuroGP-04), *LNCS*, vol. 3003, pp. 309–317. Springer-Verlag (2004)
22. Hong, J., Cho, S.: Lymphoma cancer classification using genetic programming with SNR features. In: Proc. of the European Conf. on Genetic Programming (EuroGP-04), *LNCS*, vol. 3003, pp. 78–88. Springer-Verlag (2004)
23. Howley, T., Madden, M.G.: The genetic kernel support vector machine: description and evaluation. Artificial Intelligence Review **24**(3-4), 379–395 (2005)
24. Hu, Y.: A genetic programming approach to constructive induction. In: Proc. of the 3rd Annual Conf. on Genetic Programming (GP-98), pp. 146–151. Morgan Kaufmann (1998)
25. Kishore, J., Patnaik, L., Mani, V., Agrawal, V.: Application of genetic programming for multicategory pattern classification. IEEE Transactions on Evolutionary Computation **4**(3), 242–258 (2000)
26. Koza, J.R.: Genetic Programming: on the programming of computers by the means of natural selection. The MIT Press, Massachusetts (1992)
27. Mendes, R.R.F., Voznika, F.B., Freitas, A.A., Nievola, J.C.: Discovering fuzzy classification rules with genetic programming and co-evolution. In: Proc. of the European Conf. on Principles of Data Mining and Knowledge Discovery (PKDD-01), pp. 314–325. Springer Verlag (2001)
28. Montana, D.J.: Strongly typed genetic programming. Evolutionary Computation **3**(2), 199–230 (1995)
29. Oltean, M.: Evolving evolutionary algorithms using linear genetic programming. Evolutionary Computation **13**(3), 387–410 (2005)
30. Oltean, M., Dumitrescu, D.: Evolving TSP heuristics using multi expression programming. In: Proc. of the 4th Int. Conf. on Computational Science, *LNCS*, vol. 3037, pp. 670–673. Springer (2004)
31. Papagelis, A., Kalles, D.: Breeding decision trees using evolutionary techniques. In: Proc. of the 18th Int. Conf. on Machine Learning (ICML-01), pp. 393–400. Morgan Kaufmann Publishers, San Francisco, CA, USA (2001)
32. Quinlan, J.R.: Learning logical definitions from relations. Machine Learning **5**, 239–266 (1990)
33. Quinlan, J.R.: C4.5: programs for machine learning. Morgan Kaufmann (1993)
34. Rivero, D., Dorado, J., Rabuñal, J.R., Pazos, A., Pereira, J.: Artificial neural network development by means of genetic programming with graph codification. Transactions on Engineering, Computing and Technology **16**, 209–214 (2006)
35. Rouwhorst, S., Engelbrecht, A.: Searching the forest: using decision trees as building blocks for evolutionary search in classification databases. In: Proc. of Congress on Evolutionary Computation (CEC-00). IEEE Press (2000)
36. Ryan, M., Rayward-Smith, V.: The evolution of decision trees. In: Proc. of the 3rd Annual Conference on Genetic Programming(GP-98), pp. 350–358. Morgan Kaufmann (1998)
37. Smart, W., Zhang, M.: Using genetic programming for multiclass classification by simultaneously solving component binary classification problems. In: Proc. of the European Conf. on Genetic Programming (EuroGP-05), *LNCS*, vol. 3447, pp. 227–239. Springer-Verlag (2005)
38. Suyama, A., Negishi, N., Yamaguchi, T.: CAMLET: A platform for automatic composition of inductive learning systems using ontologies. In: Proc. of the Pacific Rim Int. Conf. on Artificial Intelligence, pp. 205–215 (1998)

39. Whitley, D.: Permutations. In: T. Back, D. Fogel, T. Michalewicz (eds.) Evolutionary Computation 1: Basic Algorithms and Operators, pp. 274–284. Institute of Physics Publishing (2000)
40. Witten, I.H., Frank, E.: Data Mining: Practical Machine Learning Tools and Techniques with Java Implementations, 2nd edn. Morgan Kaufmann (2005)
41. Wong, M.L.: An adaptive knowledge-acquisition system using generic genetic programming. Expert Systems with Applications **15**(1), 47–58 (1998)
42. Wong, M.L., Leung, K.S.: Data Mining Using Grammar-Based Genetic Programming and Applications. Kluwer, Norwell, MA, USA (2000)
43. Yao, X.: Evolving artificial neural networks. Proceedings of the IEEE **87**(9), 1423–1447 (1999)

Chapter 5
Automating the Design of Rule Induction Algorithms

5.1 Introduction

Evolutionary algorithms have been used for evolving rule sets as a kind of classification model for a long time [10, 13, 15, 23]. As explained in Chapter 4, from rule-based classification models researchers turned to evolving components of data mining algorithms, and recently we have shown we can go one step further, evolving complete rule induction algorithms [18]. This latter idea is the main topic of this chapter, which introduces a Grammar-based Genetic Programming (GGP) system created to automatically evolve rule induction *algorithms*. As it will be shown later, the proposed GGP system can be used to generate robust rule induction algorithms or algorithms with performance tailored to a specific application domain (dataset). The type of the rule induction algorithms generated depends only on the data with which the GGP system is trained [16, 17, 18].

Recall that there are two main approaches that can be followed by a GGP system. The GGP system proposed here follows the solution-encoding individual approach (see Section 3.6.2), where the grammar is used to create the individuals in the GGP system's initial population, instead of taking part in a genotype-phenotype mapping process (production-rule-sequence-encoding individual approach). Figure 5.1 shows the scheme of the proposed GGP system.

The grammar contains background knowledge about the basic structure of rule induction algorithms following the sequential covering approach, and is described in Section 5.2. Each individual in the GGP system's population represents a new rule induction algorithm, potentially as complex as well-known algorithms such as CN2 [5] or PRISM [4], as explained in Section 5.3. These individuals are built by following a set of derivation steps of the grammar, as detailed in Section 5.4. The individuals (rule induction algorithms) are evaluated using a set of datasets, named meta-training set. The classification accuracies obtained from the runs of the rule induction algorithms in all the datasets in the meta-training set are used to generate the values of the fitness function for the individuals, as described in Section 5.5.

G.L. Pappa, A.A. Freitas, *Automating the Design of Data Mining Algorithms*,
Natural Computing Series, DOI 10.1007/978-3-642-02541-9_5,

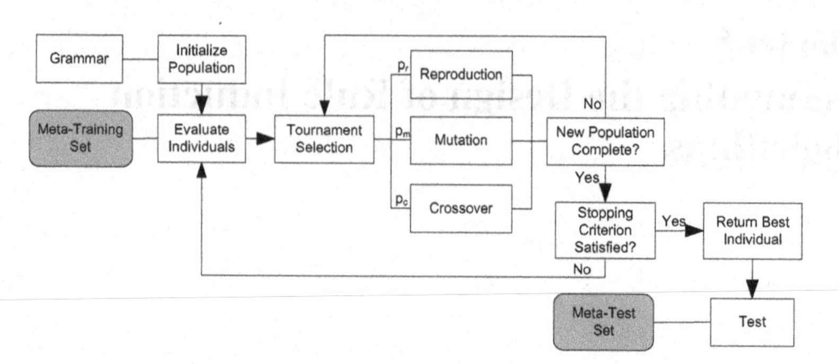

Fig. 5.1 Scheme of the grammar-based GP system for evolving rule induction algorithms

Following evaluation, an elitist reproduction strategy [12] selects the individual with the best fitness value and passes it onto the new population without any modifications. Next, a tournament selection scheme with $k = 2$ (for details, see Section 3.5.3) is used to select the individuals that will produce the new population. After selection, the winners of the tournaments undergo either reproduction, mutation, or crossover operations, depending on user-defined rates, as described in Section 5.6.

The evolutionary process is performed until a maximum number of generations is reached. At the end of the evolutionary process, the best individual (the one with the highest fitness produced along the run of the system) is returned as the solution for the problem. The chosen rule induction algorithm is then evaluated in a new set of datasets, named meta-test set, which contains datasets different from the datasets in the meta-training set.

It is important to emphasize that, to the best of our knowledge, there is no significant research that indicates that either the solution-encoding individual approach or the production-rule-sequence-encoding individual approach for grammar-based genetic programming is superior to the other. Hence, in the absence of significant evidence in favor of any of these two approaches with respect to their effectiveness, we chose to use the solution-encoding-individual approach. This choice was based on the fact that, as this approach lacks a genotype-phenotype mapping process, there is no need to worry about how effective the mapping is or how large the degree of epistasis (interaction among genes) at the genotype level is. However, we make no claim that the solution-encoding individual approach is superior for our problem domain. Performing computational experiments comparing the relative effectiveness of this approach and the production-rule-sequence-encoding approach is a topic left for further research.

Table 5.1 The grammar used by the GGP system

```
1-<Start> ::= (<CreateRuleSet>|<CreateRuleList>)[<PostProcess>]
2-<CreateRuleSet> ::= forEachClass <whileLoop> endFor
                      <RuleSetTest>
3-<CreateRuleList> ::= <whileLoop> <RuleListTest>
4-<whileLoop>::= while <condWhile> <CreateOneRule> endWhile
5-<condWhile>::= uncoveredNotEmpty |uncoveredGreater
                 (10| 20| 90%| 95%| 97%| 99%) trainEx
6-<RuleSetTest> ::= lsContent |confidenceLaplace
7-<RuleListTest>::= appendRule | prependRule
8-<CreateOneRule>::= <InitializeRule> <innerWhile>
                     [<PrePruneRule>][<RuleStoppingCriterion>]
9-<InitializeRule> ::= emptyRule| randomExample|
                       typicalExample | <MakeFirstRule>
10-<MakeFirstRule> ::= NumCond1| NumCond2| NumCond3| NumCond4
11-<innerWhile> ::= while (candNotEmpty| negNotCovered)
                    <FindRule> endWhile
12-<FindRule> ::= (<RefineRule>|<innerIf>) <EvaluateRule>
                  [<StoppingCriterion>] <SelectCandidateRules>
13-<innerIf> ::= if <condIf> then <RefineRule>
                 else <RefineRule>
14-<condIf> ::=  <condIfExamples> | <condIfRule>
15-<condIfRule> ::= ruleSizeSmaller (2| 3| 5| 7)
16-<condIfExamples> ::= numCovExp ( >| <)(90%| 95%| 99%)
17-<RefineRule> ::= <AddCond>| <RemoveCond>
18-<AddCond> ::= Add1| Add2
19-<RemoveCond>::= Remove1| Remove2
20-<EvaluateRule>::= confidence|Laplace|infoContent|infoGain
21-<StoppingCriterion> ::= MinAccuracy (0.6| 0.7| 0.8)|
                           SignifTest (0.1| 0.05| 0.025| 0.01)
22-<SelectCandidateRules> ::= 1C| 2C| 3C| 4C| 5C| 8C| 10C
23-<PrePruneRule> ::= (1Cond| LastCond| FinalSeqCond)
                      <EvaluateRule>
24-<RuleStoppingCriterion> ::= accuracyStop (0.5|0.6| 0.7)
25-<PostProcess> ::= RemoveRule EvaluateModel|<RemoveCondRule>
26-<RemoveCondRule> ::= (1Cond| 2Cond| FinalSeq)<EvaluateRule>
```

5.2 The Grammar: Specifying the Building Blocks of Rule Induction Algorithms

The main motivation for using Grammar-based Genetic Programming (GGP) systems is to give the system background knowledge about the problem being tackled. Building a complete rule induction algorithm from scratch would be certainly a much more complex problem than using a grammar to guide the process. Another role of the grammar inside the GGP solution-encoding individual framework is to guarantee that all the individuals in the GGP system's population are valid and also respect the closure property. In order to achieve that, the grammar is used to generate the initial population of individuals, and also controls crossover and mutation operations. In this work, we aim to automatically evolve a sequential covering rule induction algorithm. Therefore, our grammar presents all the elements we find appropriate to use while building rule induction algorithms. This section shows, step by step, how the grammar was defined.

The proposed grammar is presented in Table 5.1. It uses the BNF terminology introduced in Section 3.6.1, and its *Start* symbol is represented by the nonterminal with the same name. Recall that nonterminals are wrapped into "<>" symbols, and each of them originates one or more production rules. Grammar symbols not presented between "<" and ">" are terminals. In the context of rule induction algorithms, the set of nonterminals and terminals are divided into two subsets. The first subset includes general programming elements, such as *if* conditional statements and *for* or *while* loops, while the second subset includes components directly related to rule induction algorithms, such as *RefineRule* or *PruneRule*.

The nonterminals in the grammar represent high-level operations, such as a *while* loop (*whileLoop*) or the procedure performed to refine a rule (*RefineRule*). The terminals, in turn, represent very specific operations, such as *Add1*, which adds one condition at a time to a candidate rule during the rule refinement process (*RefineRule*). Each terminal is associated with a building block. A building block represents an "atomic operation" (from the grammar's viewpoint) which does not need anymore refinements. Building blocks will be very useful during the phase of rule induction code generation, and are explained in Section 5.5.1.

As observed in Table 5.1, the grammar contains 26 nonterminals (NTs), which originate 83 production rules. The grammar was carefully created by doing a comprehensive study of the main elements of the pseudocodes of basic rule induction algorithms (Algorithms 2.2, 2.3, and 2.4 introduced in Chapter 2).

The first NT, *Start*, generates one of the two rule models described by Algs. 2.2 and 2.4: a rule set (*CreateRuleSet*) or a rule list (*CreateRuleList*). It also determines if the rule models generated will be post-processed or not.

NT 2 and NT 3 describe the nonterminals *CreateRuleSet* and *CreateRuleList*. They represent the outer loops described in Algs. 2.2 and 2.4. The *forEachClass* terminal allows rules to be built for each class in turn. The *whileLoop* (NT 4) keeps adding rules to the rule set or list until *condWhile* is satisfied.

condWhile (NT 5) is satisfied under one of the two conditions: 1) When all the examples in the training set are covered by the current set of rules; or 2) when a percentage of examples in the training set or a fixed number of them is covered by the current set of rules. We used 90%, 95%, 97%, and 99% when working with percentages, and ten or 20 examples when working with the absolute numbers. While the first condition requires that rules be produced for all the examples in the training set, the second condition gives the algorithm some flexibility, and helps avoid overfitting.

The nonterminals *RuleSetTest* (NT 6) and *RuleListTest* (NT 7) define how the rules will be applied when classifying new instances. As explained in Section 2.4.1, the rules in a decision list can be only applied in order. However, during the creation of the list, the rules can be appended or prepended to it. The standard approach is to append rules to the list, but Webb and Brkic [24] suggested that prepending rules to lists generates simpler models. They argue that there are usually simple rules that cover many of the positive examples and also a few negative examples. Leaving these simpler rules at the end of the model would allow more specific or complicated rules to handle these exceptions before the more general rules are applied.

In the case of rule sets, NT 6 defines which tiebreaking criterion will be applied in cases where two or more rules classify a test example (unseen during training) in two different classes. Among the options are the use of the LS content of a rule (see Eq. (2.4)) or the Laplace estimation (see Eq. (2.2)). Note that both these measures can also be used to evaluate rules when creating them, although LS content is not used for this purpose in the current version of the grammar.

The body of the *while* loop described by NT 4 implements the nonterminal *CreateOneRule* (NT 8), which represents the pseudocode of Alg. 2.3 discussed in Chapter 2. Rules are built following three basic steps, initialization, refinement, and selection, the last two steps being iterative.

The first step generates an initial rule. As shown in NT 9, a rule can be initialized in four different ways: (1) with an empty antecedent (represented by the terminal *emptyRule*), (2) from a seed example (picked randomly from the training set, and represented by the terminal *randomExample*), (3) from a typical [28] training example (represented by the terminal *typicalExample*) or, (4) according to the frequency of the attribute-value pairs in the dataset.

The concept of typical examples is borrowed from the instance-based learning literature [28]. An example is said to be typical if it is very similar to the other examples belonging to the same class it belongs to, and not similar to the other examples belonging to other classes. In other words, a typical example has high intra-class similarity and low inter-class similarity. Equation (5.1) shows how the typicality of an example is calculated. It is the ratio of the intra-class and inter-class similarities, where the similarity between the examples e_1 and e_2 is calculated as the complement of the distance between the examples e_1 and e_2, and P and N represent the number of positive and negative examples in the training set, respectively. (We consider here a two-class problem for simplicity, but of course this concept can be generalized to a problem with more than two classes.) The distance between the examples e_1 and e_2 is calculated as a simple Euclidean distance, as shown in Eq. (5.2). In Eq. (5.2), m represents the number of attributes in the dataset, and max_i and min_i correspond to the maximum and minimum values assumed by the attribute i in the case of continuous (real-valued) attributes. In the case of nominal (or categorical) attributes, the distance with respect to a single attribute can only be 0 (if attribute values have the same value) or 1 (if attribute values have different values). If one of the attribute values is a missing value, the difference between it and any other attribute value is set to 0.5.

$$typicality(e) = \frac{(\sum_{i=1}^{P}(1 - distance(e,i)))/P}{(\sum_{j=1}^{N}(1 - distance(e,j)))/N}, \tag{5.1}$$

where for continuous attributes

$$distance(e_1,e_2) = \sqrt{\frac{1}{m}\sum_{i=1}^{m}(\frac{e_{1i}-e_{2i}}{max_i - min_i})^2} \tag{5.2}$$

In the case of initializing a rule according to the frequency of the attribute-value pairs, the nonterminal *MakeFirstRule* (NT 10) creates a rule with 1, 2, 3, or 4 conditions in the rule antecedent. The pairs of attribute-values (conditions) inserted in the rule antecedent are selected using a probabilistic selection scheme, in which the probability of selecting a given attribute-value pair is proportional to the frequency of that attribute-value pair in the training set.

After the initial rule is created using one of the methods described above, it is set as the current best rule, and then the rule refinement process starts. It is an iterative process that occurs inside the *innerWhile* loop (NT 11), which follows the nonterminal *InitializeRule* in NT 8. As described in the conditions of the *innerWhile* loop, rules can be refined until they do not cover any negative examples, as stated by the terminal *negNotCovered*, or until a set of candidate rules is not empty (*candNotEmpty*). This set of candidate rules refers to the rules which are undergoing the refinement process. At the first iteration, the only candidate rule is the one created in the initialization process. In the remaining iterations, they are the rules selected by the nonterminal *SelectCandidateRules* (NT 22), as will be explained later.

At each iteration of the *innerWhile*, the nonterminal *FindRule* (NT 12) creates a rule by (a) finding all the possible refinements of the initial rule (*RefineRule*); (b) evaluating the rules generated through the refinements (*EvaluateRule*); and (c) selecting a fixed number of created rules to keep performing the refinement process. Moreover, *FindRule* also defines alternative ways to refine the rules (*innerIf*) and allows the definition of an alternative criterion to stop the rule refinement process (*StoppingCriterion*).

RefineRule (NT 17) changes the current candidate rules by adding (*AddCond*, NT 18) or removing (*RemoveCond*, NT 19) conditions (attribute-value pairs) to or from them. According to the production rules generated by NT 18 and NT 19, either one or two conditions at a time can be inserted into or removed from a rule.

Most of the current rule induction algorithms, as described in Section 2.4.2, use a top-down or a bottom-up search while looking for rules. Only a few of them, like SWAP-1 [25], implement a bidirectional search strategy. However, a bidirectional search might allow us to fix up rules by removing or adding conditions from or to it (especially in the case of greedy searches, where only one candidate rule goes through the refinement process). In the grammar presented in Table 5.1, the nonterminal *innerIf* (NT 13) changes the way the rules are refined according to their size (*condIfRule*, NT 15) or the number of examples covered by the current rule set or list (*condIfExamples*, NT 16). In the former case, rules having size (number of conditions) smaller than 2, 3, 5, or 7 might be refined in a different way than rules having size greater than or equal to 2, 3, 5, or 7. In the latter case, a decision is made based on the percentage of examples of the training set covered by the current rule set or list. It considers whether the number of covered training examples is smaller or greater than 90%, 95%, or 99% of the total number of examples in the training set.

The motivation to include this *innerIf* statement in the grammar is that the algorithm can choose, for example, to add two conditions at a time to the rule while its size is small (i.e., the number of examples covered by the rule is hopefully large enough to detect attribute interaction). However, as the rule size grows, and the number of examples covered by the rule shrinks, it might be easier to improve its predictive power just by adding one condition at a time instead of two. The same argument is valid for the number of examples in the training set covered by the rule set or list. The fewer the examples left uncovered, the more difficult to intuitively find a combination of conditions that would improve the current rule. However, even though our intuition says this might be the most appropriate way to use the *innerIf* included into the grammar, the GGP system might come up with other counterintuitive but more effective way of using it.

The rules created through the refinement process are evaluated using one of the measures defined in NT 20, namely confidence, Laplace estimation, information content, or information gain (previously defined in Section 2.4.3). These measures take into account the number of positive and negative examples the rule covers, and they will be used in the last phase of the rule generation process to select a subset of rules to go through further refinements.

During the rule selection phase, the rule evaluation function can be combined with some other criterion in order to select the best rules found so far. Hence, in some cases, a rule has also to fulfill a refinement stop criterion imposed by the optional nonterminal *StoppingCriterion* (NT 21). *StoppingCriterion* requires a rule to have a minimum accuracy or to be significant according to a statistical significance test during the rule selection process.

The minimum accuracy criterion calculates the accuracy (or confidence, as described in Eq. (2.1)) of the rule and compares it with a threshold. The statistical significance test – a χ^2 (chi-squared) test – calculates the distance between the distribution of the classes of the examples covered by the rule and the expected distribution (given by the frequencies of examples in each class for the entire training set). The lower the value of the distance, the higher the probability that the concept represented by the rule is due to chance [6]. The numbers 0.6, 0.7, and 0.8 following the terminal *MinAccuracy* represent the thresholds used for accuracy, and the figures 0.1, 0.05, 0.025, and 0.01 after the terminal *SignifTest* are confidence levels for the significance test.

The rule selection process, based on the values of the evaluation function and stopping criterion, selects a number of candidate rules to enter the next iteration of the rule refinement process. As specified in NT 22, the number of selected rules can vary from 1 to 5, or it can be 8 or 10. Regardless of the number of selected rules, the current best rule is replaced only if its evaluation function value is worse than the value of the best selected rule.

When this single rule building process terminates, a new rule is added to the rule set or list being produced. But before this happens, a last operation can be performed as a result of applying NT 8: rule pre-pruning. *PrePruneRule* (NT 23) implements a pre-pruning method that tries to simplify a rule by removing one condition or a set of conditions from its antecedent. Rules can be simplified in three ways: (1)

Removing one condition at a time (*1Cond*) from its antecedent, as long as the new rule is better than the original rule according to an evaluation criterion; (2) removing the last added condition (*LastCond*) from its antecedent; or (3) removing a sequence of conditions from the end of the rule antecedent (*FinalSeqCond*), as long as the new rule is better than the original rule according to an evaluation criterion. During the pre-pruning phase, rules are evaluated in a set of data different from the one used to build them.

Recall that rules are created until all or a large part of the examples in the training set are covered by the generated rules. However, the process of rule creation can also be halted when a condition defined by the nonterminal *RuleStoppingCriterion* (NT 24) is not satisfied. This condition is usually based on some property of the last rule found, and includes verifying if the accuracy (confidence) of the just-produced rule is greater than a threshold. We use threshold values[1] of 0.5, 0.6, and 0.7.

Once the rule set or list is completed, the rule induction algorithm can still perform a last operation: post-process the rule model. The presence of a post-process step in the algorithm is determined by the application of NT 1. Post-processing methods (NT 25) can apply the same techniques used to pre-prune a single rule to all the rules in the rule model. Hence, *RemoveCondRule* (NT 26) describes similar methods to the ones described in NT 23. After the model is completed, it can be simplified by removing one (*1Cond*) or two (*2Cond*) conditions at a time from the rule antecedents. In the case of rule sets, the model is simplified as long as the new rule is better than the original one according to an evaluation criterion. In the case of rule lists, this process goes on while the accuracy of the entire model is not reduced. The option of removing a final sequence of conditions is also available while post-processing rules. Besides, more compact models can be tried out by removing one by one entire rules from the current model, as implemented by the *RemoveRule* terminal on the right-hand side of the production rule generated by the NT 25. After each rule is removed from the rule set the whole model has to be reevaluated, as indicated by the terminal *EvaluateModel*. Again, as in the pre-pruning phase, the rules or rule sets being post-processed are evaluated in a set of data different from the one used to build them.

By applying the production rules defined by the grammar, we can generate up to approximately five billion different rule induction algorithms [16, p. 196]. Each of these rule induction algorithms can be represented by an individual in the GGP system's population.

5.2.1 The New Rule Induction Algorithmic Components in the Grammar

As explained before, the grammar presented in Table 5.1 contains knowledge about how humans design rule induction algorithms, but it also presents some new compo-

[1] The accuracy is normalized to return a number in [0...1].

nents that, to the best of our knowledge, were not used in rule induction algorithms before.

The major new components inserted into the grammar are

- The terminal *typicalExample*, which creates a new rule using the concept of typicality, borrowed from the instance-based learning literature.
- The nonterminal *MakeFirstRule*, which allows the first rule to be initialized with one, two, three, or four attribute-value pairs, selected probabilistically from the training data in proportion to their frequency. Attribute-value pairs are selected subject to the restriction that they involve different attributes (to prevent inconsistent rules such as "sex = male AND sex = female").
- The nonterminal *innerIf*, which allows rules to be refined in different ways (e.g. adding or removing one or two conditions at a time to or from the rule) according to the number of conditions they have or the number of examples the rule set or list covers.
- Although some methods use rule look-ahead, i.e., they insert more than one condition at a time into a set of candidate rules, we did not find in the literature any rule induction algorithm which removes two conditions at a time from a rule. This is implemented by the terminal *Remove2*. See [11] for a more detailed discussion on bottom-up look-ahead algorithms.

Note that the list above shows a set of single components which are new "building blocks" of rule induction algorithms. These components increase the diversity of the candidate rule induction algorithms considerably, but it is the combination of the "standard" and new components which will potentially contribute to the creation of a new rule induction algorithm different from the conventional algorithms.

5.3 Individual Representation

As pointed out before, the GGP system described in this chapter follows the solution-encoding individual approach, where each individual is represented by a derivation tree created by applying a set of production rules from the grammar.

Figure 5.2 shows an example of a GGP system's individual. The root of the tree is the nonterminal *Start*. The tree is then derived by the application of production rules for each nonterminal. For example, *Start* (NT 1) generates the nonterminal *CreateRuleList* (NT 3), which in turn produces the nonterminals *whileLoop* and *RuleListTest*. This process is repeated until all the leaf nodes of the tree are terminals.

Every time the application of a production rule involves an option between two or more symbols on the right-hand side of a production rule, each of the candidate symbols has the same probability of being chosen. Hence, the generation of the derivation tree representing an individual is a nondeterministic process, as is usual in GP.

Recall that an individual represents a complete rule induction algorithm. Thus, in order to extract from the tree the pseudocode of the corresponding rule induction algorithm, we read all the terminals (leaf nodes) in the tree from left to right. The tree

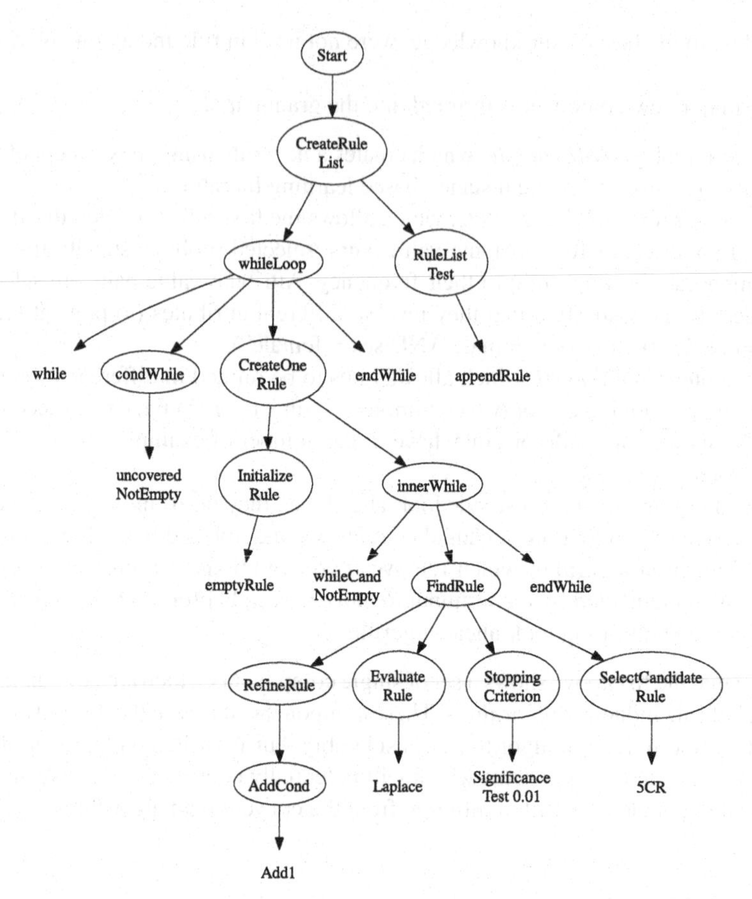

Fig. 5.2 Example of a GGP sytem's individual

in Fig. 5.2, for example, represents the pseudocode described in Alg. 5.1, expressed at a high level of abstraction. This algorithm actually represents an instance of the well-known CN2 algorithm [5] producing an ordered list of rules, with the beam width (or star size, using the CN2 terminology) parameter set to 5 and the statistical significant test threshold set to 0.01.

5.4 Population Initialization

The GGP system proposed in this work generates the initial population using a method similar to the one suggested by Whigham [26]. It starts the population generation procedure by calculating the minimum tree depths for all the production rules of the grammar.

The minimum tree depth is used to guide the selection of production rules in the grammar in a way that programs with different depths are generated. Production

Algorithm 5.1: Rule induction algorithm represented by the derivation tree in Fig. 5.2

```
RuleList = ∅
repeat
    bestRule = an empty rule
    candidateRules = ∅
    candidateRules = candidateRules ∪ bestRule
    while candidateRules ≠ ∅ do
        newCandidateRules = ∅
        for each candidateRule CR do
            Add 1 condition at a time to CR
            Evaluate CR using the Laplace estimation
            if CR is significant at the 0.01 significance level then
                newCandidateRules = newCandidateRules ∪ CR
                if CR is better than bestRule then
                    bestRule = CR
        candidateRules = 5 best rules selected from newCandidateRules
    RuleList = RuleList ∪ bestRule
    Remove from the training set all examples covered by bestRule
until all examples in the training set are covered
```

rules which generate only terminals have minimum tree depth of 1. The minimum depth of other production rules is calculated in a bottom-up fashion. Consider the production rule <A> ::= <B> <C>. If the minimum depth of <B> and <C> are known, the minimum depth of <A> is set as the maximum of the values of the minimum depths of <B> and <C> + 1. If the values of <B> and/or <C> are unknown, their minimum tree depths are calculated first, in a recursive manner.

The depth of the GGP system's trees does not vary a lot in the grammar presented in Table 5.1. The minimum depth of a tree is 9, and the maximum depth is 10. This is because the current grammar does not have any recursive production rules (i.e., productions like <A> ::= <B> <A> | c). Very different individuals will greatly vary in the number and contents of tree nodes, but not in tree depth. In the current version of the system, half of the individuals in the first population are created with depth 9, and the other half with a tree depth of 10. However, the system does support recursive production rules, and in this case an equal number of individuals for each depth – with depths varying from the minimum depth of the *Start* symbol to the maximum depth parameter set by the user – would be generated. The system guarantees that all the individuals in the first population are different from each other.

An issue that has to be given attention during the population initialization process is that some combinations of nonterminals and terminals in the grammar may generate individuals that are semantically invalid. By semantically invalid individuals we mean that the rule induction algorithms generated by these individuals will, most of the time, execute infinite loops. Figure 5.3 shows an example of part of a simple rule induction algorithm defined by a GGP system's individual that would lead to an infinite loop.

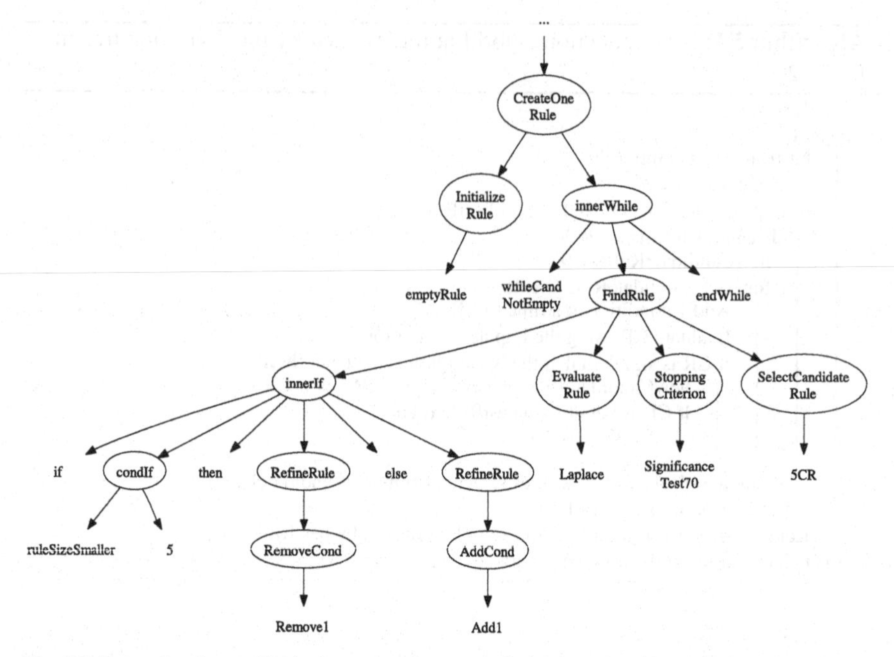

Fig. 5.3 Example of an individual representing a rule induction algorithm which will lead to an infinite loop

The individual represented in Fig. 5.3 starts producing a rule with an empty condition, and removes one condition at a time from it as long as the size (number of conditions) of the produced rule is 5. As the first rule is empty, the *if* part of the *if* statement does not do anything, and the *else* part is unreachable. Because the set of candidate rules will never change, i.e., it will always have the empty rule, the condition to finish the *innerWhile* loop will never be satisfied.

Some infinite loop situations, such as the one presented in Fig. 5.3, can be predicted by a careful evaluation of the possible combinations of nonterminals and terminals in the grammar, and are set up as constraints during the population initialization process and breeding operations. The current version of the GGP system imposes the following constraints when creating individuals:

1. The nonterminal *AddCond* cannot be combined with the terminals *randomExample* and *typicalExample*, which may be generated by the nonterminal *InitializeRule*. This is because it does not make sense to add more conditions to a rule represented by a complete example taken from the training set.
2. Using the same rationale, the nonterminal *RemoveCond* cannot be used in association with the terminal *emptyRule* generated by the nonterminal *AddCond*, or the terminals *numCond1* and *numCond2* generated by the nonterminal *MakeFirstRule*.
3. The nonterminal *innerIf* has to incorporate the constraints imposed by item 1 so that situations like the one presented in Fig. 5.3 are avoided. The following combinations are not legal in an individual using an *innerIf*:

- When the nonterminals *condIfRule* and *RemoveCond* (in the *if* part of the *if* statement) are used together, they cannot be combined with the terminals *emptyRule*, *numCond1*, or *numCond2*.
- When the nonterminals *condIfRule* and *AddCond* (in the *else* part of the *if* statement) are used together, they cannot be combined with the terminals *randomExample* or *typicalExample*.
- When the terminal > (<), generated by the nonterminal *condIfExamples*, is combined with *RemoveRule* (in the *else* (*if*) part of the *if* statement), it cannot be combined with the terminals *emptyRule*, *numCond1*, or *numCond2*.
- When the terminal > (<), generated by the nonterminal *condIfExamples*, is combined with *AddRule* (in the *else* (*if*) part of the *if* statement), it cannot be combined with the terminals *randomExample* or *typicalExample*.

Note that these constraints avoid the most general cases of infinite loops, but we cannot guarantee they avoid all of them. Hence, individuals generating infinite loops are penalized during the fitness evaluation process, as explained in Section 5.5, and are unlikely to survive for many generations.

In any case, evaluations of the method used to generate the initial population showed that the constraints do decrease significantly the number of individuals executing infinite loops. Furthermore, the population initialization process guarantees diversity in the population, as can be observed in the graphs shown in Figs. 5.4 through 5.6. These graphs show the distribution of some terminals during the evolution of the GGP system. In Fig. 5.4, for instance, we can observe how the terminals generated by *InitializeRule* were distributed. Note that *InitializeRule* generates the terminals *emptyRule*, *randomExample*, and *typicalExample* and the nonterminal *<MakeFirstRule>*, which in turn generates the terminals *NumCond1*, *NumCond2*, *NumCond3*, and *NumCond4*.

As can be observed, there is always a balanced distribution of these terminals in the first generation. Figure 5.5 shows a good example of how a symbol that starts dominating a population can be replaced by another during the evolution. This is what happens with *Add1* and *Add2*. However, in the case of this particular graph, individuals containing *if* statements can present both terminals, *Add1* and *Add2*. In Fig. 5.6 we can also observe how the *accuracy* measure initially is more used than the *Laplace estimation* measure, but at a certain stage in evolution this situation is reversed.

5.5 Individual Evaluation

An evolutionary algorithm works by selecting the fittest individuals of a population to reproduce and generate new offspring. Individuals are selected based on how good their corresponding candidate solutions are to solve the problem being tackled. In our case, we need to evaluate how good a rule induction algorithm is.

In the rule induction algorithm literature, comparing different classification algorithms is not a straightforward process. There is a variety of metrics that can

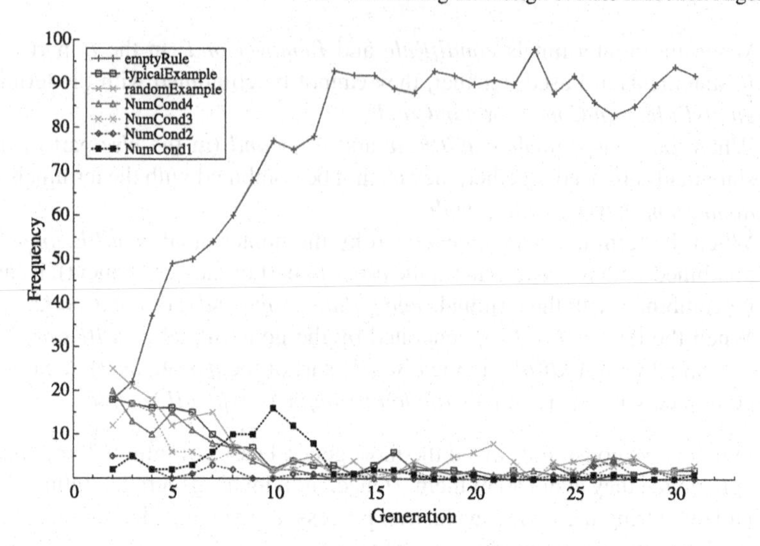

Fig. 5.4 Frequencies of terminals responsible for the rule induction algorithm rule initialization process during the GGP system's evolution

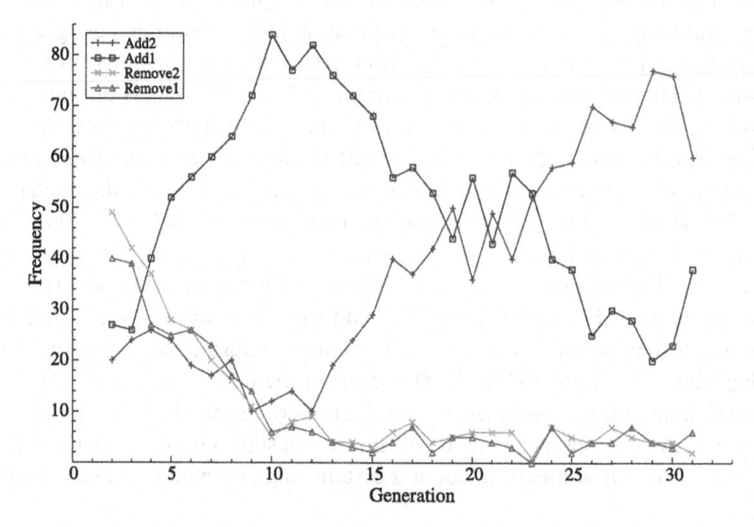

Fig. 5.5 Frequencies of terminals responsible for the rule induction algorithm rule refinement process during the GGP system's evolution

be used to estimate how good a classifier is, including classification accuracy, sensitivity and specificity, and ROC analysis [8]. There are studies comparing these different metrics and showing advantages and disadvantages in using each of them [2, 9]. Nevertheless, as pointed out by Caruana and Niculescu-Mizil [2], in supervised learning there is one ideal model, and "we do not have performance metrics

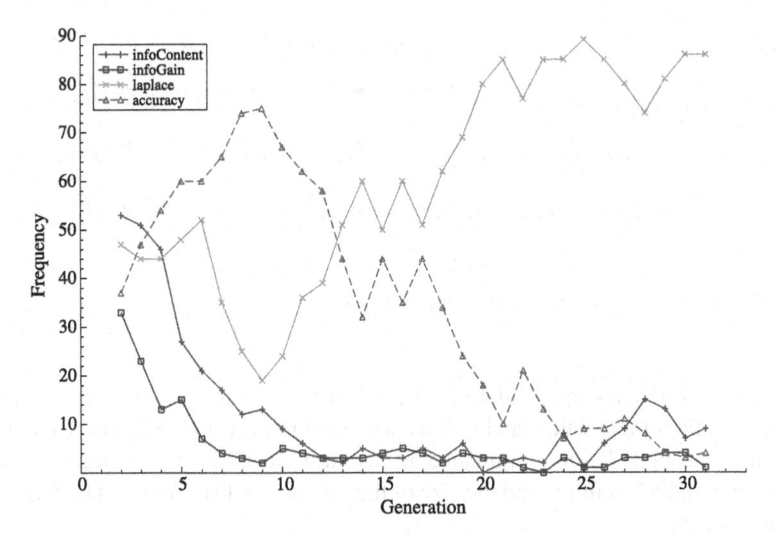

Fig. 5.6 Frequencies of terminals responsible for the rule induction algorithm rule evaluation process during the GGP system's evolution

that will reliably assign best performance to the probabilistic true model given finite validation data."

Classification accuracy is still the most common metric used to compare classifiers, although some authors have shown the pitfalls of using classification accuracy when evaluating induction algorithms [21], especially because it assumes equal misclassification costs and known class distributions, and others have introduced ROC analysis as a more robust accuracy measure. Based on the idea of using a simpler measure when first evaluating the individuals produced by the GGP system, we chose to use a measure based on standard classification accuracy to compose the fitness of the GGP system.

In this framework, a rule induction algorithm RI_A is said to outperform a rule induction algorithm RI_B if RI_A has better classification accuracy in a set of classification problems. Thus, in order to evaluate the rule induction algorithms being evolved, we selected a set of classification problems and created a meta-training set. The meta-training set consists of a set of datasets, each of them divided as usual into training and validation sets.

As illustrated in Fig. 5.7, each individual in the GGP system's population is decoded into a rule induction algorithm using a GGP/Java interface, as will be detailed in Section 5.5.1. The Java code is then compiled, and the resulting rule induction algorithm run in all the datasets belonging to the meta-training set. It is a conventional run where, for each dataset, a set or list of rules is built using the set of training examples and evaluated using the set of validation examples. After the rule induction algorithm is run in all the datasets in the meta-training set, the accuracy in the validation set and the rule sets or lists produced for all datasets are returned. These two measures can be used to calculate a fitness function. In this work, we investi-

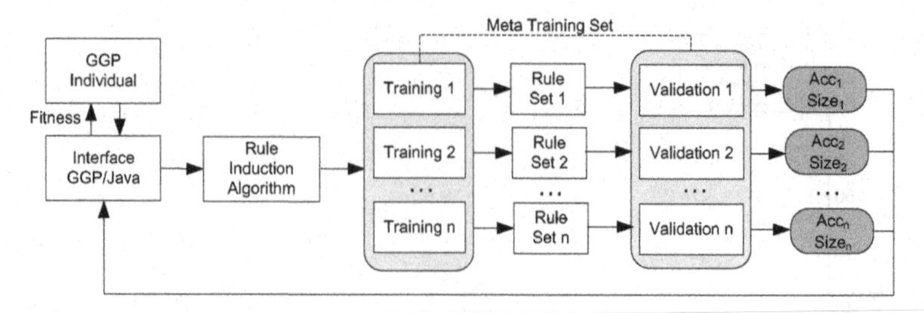

Fig. 5.7 Fitness evaluation process of a GGP system's Individual

gated two approaches to calculate the fitness. The first one uses only the accuracy to generate the fitness value, and will be described in Section 5.5.2. The second one uses the concept of Pareto dominance discussed in Section 3.5.2 to simultaneously optimize both the accuracy and the size of the rule sets or lists, and will be described in Section 5.5.3.

5.5.1 From a Derivation Tree to Java Code

In the process of evaluating a rule induction algorithm, the derivation tree initially created by using the production rules of the grammar has to be converted into real machine code, which can be executed in a set of classification problems and generate rule-based classification models for the corresponding datasets. This process of conversion of the GGP system's trees into code is the focus of this section, and is illustrated in Fig. 5.8.

The first thing to point out in Fig. 5.8 is that each terminal in the grammar is associated with Java code, which is an implementation of the building block represented by the associated terminal. Recall that each terminal in the grammar represents a building block, that is, an "atomic" operation (at a high level of abstraction, from the point of view of the grammar) performed by a rule induction algorithm. For instance, the terminal *Add1* is associated with the building-block described in Alg. 5.2. This building block is implemented as the Java method *public List add1Condition(Rule r)*, which takes the current rule as a parameter and returns a list of possible rule refinements. The system was implemented in Java because, instead of implementing the whole system from scratch, we used WEKA [27], an open source data mining tool written in Java, to speed up the implementation process.

The building blocks (Java code) associated with the terminals of the grammar were implemented in two phases, with the purpose of quickly obtaining a first set of results to validate the proposed idea. In the first phase, the grammar terminals dealt only with nominal attributes. In the second phase, their implementation was extended to also handle numerical attributes. The terminals whose implementation went through major extensions in this second phase were the ones responsible for refining rules by adding or removing conditions to or from it. After the completion

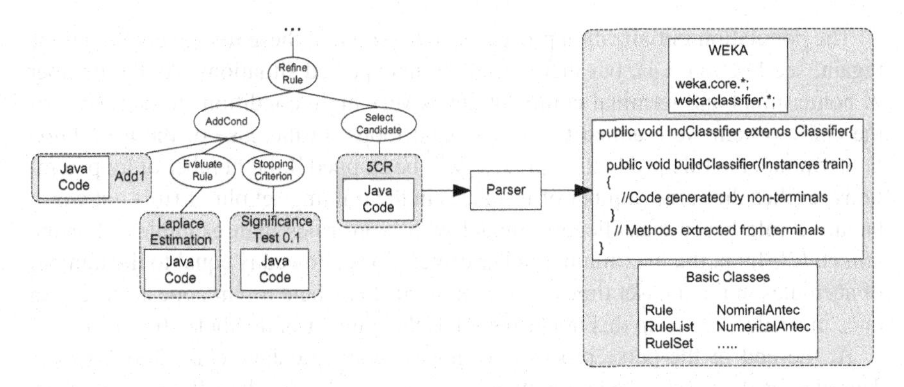

Fig. 5.8 Example of the method used to extract Java code from the GGP system's individuals

Algorithm 5.2: AddCond(Rule R)

```
refinements = ∅
for i = 0 to i < number of attributes A in the training set do
    for j = 0 to j < number of values V that A_i can take do
        newRule = R ∪ (A_i,V_j)
        refinements = refinements ∪ newRule
return refinements
```

of this second phase, the grammar became flexible enough to produce rule induction algorithms that represent rule conditions of the form <*attribute*, *operator*, *value*>, where *operator* is "=" in the case of nominal attributes and is "$\geq$" or "$\leq$" in the case of numerical attributes.

The approach followed to generate numerical conditions is similar to the one implemented by the Ripper [7] and C4.5 [22] algorithms, where the values of a numerical attribute are sorted, and all threshold values considered. The best threshold value is chosen according to the information gain associated with that attribute-value pair.

As observed in Fig. 5.8, there is a parser at the center of the process converting a GGP-tree to its correspondent Java code. The parser reads the symbols in the tree and inserts the code associated with them into a Java class named *IndClassifier*. When reading a nonterminal in the tree, the parser might add to the method *buildClassifier* of the class *IndClassifier* either (1) some standard code or (2) a call for a method defined by a terminal, or take no action. We call standard code the code that is usually found in all rule induction algorithms that contain that nonterminal, despite the terminals it generates. When reading a terminal, the parser adds to the class *IndClassifier* the methods associated with that terminal.

During the creation of the rule induction algorithm, there is another point that has to be taken into account: infinite loops. As explained in Section 5.4, although all the methods associated with a terminal perform the task they are supposed to, combinations of some symbols of the grammar may lead to infinite loops.

The population initialization process handles some of these invalid combinations (again, see Section 5.4), but there might be unexpected situations. As the number of nonterminal and terminal combinations is very high, conditions to halt a run of the GGP system were added to its two main loops. In the case of the *whileLoop* nonterminal, the rule production process will be stopped if the number of loop iterations reaches the same number of instances in the training set plus 1 (it is not worth having a rule-based classification model with more rules than examples). For the *innerWhile* loop, the maximum number of iterations allowed is equal to the number of attributes in the dataset times 2 (as a numerical attribute might appear more than once in a rule, we set up this upper bound as the number of attributes times 2).

At the end of the parse process, we have a complete Java class *IndClassifier*. This class is then associated with other basic Java classes (such as *Rule*, *Antecedent*, *RuleList*, and *RuleSet*) and some of the WEKA packages, allowing us to run the generated rule induction algorithm in a set of classification problems. Note that although the original idea was to use as much of the WEKA code as possible, during the development of the system we realized that the implementation of the classifiers in WEKA was not modular enough to make the automatic generation of the rule induction algorithms' code simple. Hence, the system does not use the methods implemented for specific classifiers in WEKA, but only its *core* package (which deals with the basic operations of any classifier, like reading the examples and generating data statistics) and some of the classes in the package *classifier*.

5.5.2 Single-Objective Fitness

As shown in Fig. 5.7, once we convert the GGP system's trees into rule induction algorithm Java code, we run them in a set of classification problems. For each classification problem (i.e., each dataset in the meta-training set), we obtain a rule model from the training set and an accuracy rate from the validation set, and then we need to summarize this information into a straightforward fitness value.

The most intuitive way of performing this summarization, which was first used to evaluate the proposed GGP system, was to use the average of the accuracies of the GGP-RI (the Rule Induction algorithm evolved by the GGP system) over all classification problems as the fitness value. However, analyzing the results obtained in the first runs of the GGP system with this fitness function, we realized it was not a good one. There is a simple explanation for that. The GGP system is calculating an average over different datasets with very different baseline accuracies (i.e., accuracies obtained when using the class of the majority of the training examples to classify new examples in the validation set). For instance, let us assume we have a dataset DS_1 with baseline accuracy 90% and a dataset DS_2 with baseline accuracy 60%. The GGP system creates two rule induction algorithms, GGP-RI_1 and GGP-RI_2. GGP-RI_1 obtains accuracies of 92% and 70% for DS_1 and DS_2, respectively; GGP-RI_2 obtains accuracies of 100% and 62% for DS_1 and DS_2, respectively. The average of the accuracies of GGP-RI_1 is the same as the average of the accuracies of GGP-RI_2, which is 81%. However, improving the accuracy from 90% to 100% is much more difficult than increasing it from 60% to 70%.

It was clear that the accuracy itself was not a good measure of how much better GGP-RI_1 was compared to GGP-RI_2. Hence, we tried two other fitness functions. The first one was based on the geometric average over the values of the function $sens_i \times specif_i$ (described in Eq. (5.3), combining sensitivity and specificity) for each dataset i in the meta-training set. Equation (5.3) describes the $n_{classes}$-root of the product of the sensitivity (true positive rate) and the specificity (true negative rate) for all classes of a dataset. Recall that the TP, FP, TN, and FN terms stand for the number of true positives (positive examples correctly classified as positive), false positives (negative examples wrongly classified as positive), true negatives (negative examples correctly classified as negative), and false negatives (positive examples wrongly classified as negative). In the case of datasets with more than two classes, when calculating the values of TP, FP, TN, and FN, each class in turn is considered as the positive class, and the remaining classes as the negative class.

$$sens_i \times specif_i = \sqrt[nClasses]{\prod_{i=1}^{nClasses} \frac{TP_i}{TP_i + FN_i} \times \frac{TN_i}{TN_i + FP_i}} \quad (5.3)$$

The second fitness function was implemented as the arithmetic average of the values of function fit_i (defined in Eq. (5.4)) for each dataset i in the meta-training set. In the definition of fit_i given in Eq. (5.4), Acc_i represents the accuracy (on the validation set) obtained by the rules discovered by the rule induction algorithm in dataset i. $DefAcc_i$ represents the default accuracy (the accuracy obtained when using the class of the majority of the training examples to classify new examples in the validation set) in dataset i.

$$fit_i = \begin{cases} \frac{Acc_i - DefAcc_i}{1 - DefAcc_i}, & \text{if } \mathrm{Acc}_i > \mathrm{DefAcc}_i \\ \frac{Acc_i - DefAcc_i}{DefAcc_i}, & \text{otherwise} \end{cases} \quad (5.4)$$

According to the definition of fit_i, if the accuracy obtained by the classifier is better than the default accuracy, the improvement over the default accuracy is normalized by dividing the absolute value of the improvement by the maximum possible improvement. In the case of a drop in the accuracy with respect to the default accuracy, this difference is normalized by dividing the negative value of the difference by the maximum possible drop (the value of $DefAcc_i$).

Hence, fit_i returns a value between -1 (when $Acc_i = 0$) and 1 (when $Acc_i = 1$). As explained before, the degree of difficulty of the classification task depends strongly on the value of $DefAcc_i$. The above fitness function recognizes this and returns a positive value of fit_i when $Acc_i > DefAcc_i$. For instance, if $DefAcc_i = 0.95$, then $Acc_i = 0.90$ would lead to a negative value of fit_i, as it should.

A set of experiments was performed using both the fitness functions defined in Equations (5.3) and (5.4). The graph in Fig. 5.9 shows the average values for both the fitness functions defined in Equations (5.3) and (5.4) over all the 100 individuals of a

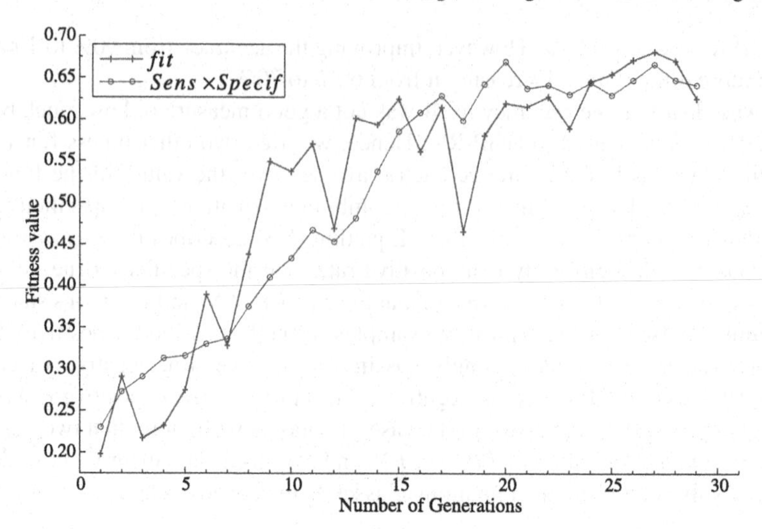

Fig. 5.9 Comparing the average fitness values given by fit and $sens \times specif$ in a GGP system's run with 100 individuals and 30 generations

typical run of the GGP system evolved for 30 generations. The graph was produced in an experiment where the meta-training set contained the datasets *monks-2*, *monks-3*, *balance-scale*, *lymph*, and *zoo*, obtained from [14]. As we can observe, the curve generated by the average value of $sens_i \times specif_i$ ($Sensitivity \times Specificity$) over all datasets i is smoother than the average fit_i curve (*fit*), but in general the behavior of both measures is similar. While the improvement of fit along the generations grows faster than the $sens \times specif$, the same is true for the declines.

Although the graph shows that in general the behavior of the two measures is consistent, the rule induction algorithms evolved by GGP system using fit were more robust when evaluated in a meta-test set containing datasets different from the ones used in the meta-training set. Hence, based on these empirical results, the average value of fit_i over all the datasets (as defined in Eq. (5.5)) was chosen as the current GGP system's fitness function. More precisely, the fitness is computed as

$$fitness = \sum_{i=1}^{nDataSets} \frac{fit_i}{nDataSets}, \tag{5.5}$$

where *nDataSets* is the number of datasets in the meta-training set.

Recall that some of the individuals generated by the GGP system can generate rule induction algorithms whose runs will enter into infinite loops. These programs have to be penalized by the fitness function, so they have a very small chance of breeding. We chose to penalize programs that enter into infinite loops in a particular dataset by setting their fitness in that dataset to *minus* the number of datasets in the meta-training set. In this way, the fitness value in that dataset has a big impact on the value of the average fitness over all the datasets in the meta-training set.

When analyzing the results of these preliminary experiments to choose the most appropriate fitness function to guide the GGP system, we noticed three facts: (1) The values of the fitness obtained by the individuals when evaluating them in the meta-training set were much superior than the ones obtained when running the evolved rule induction algorithms in the meta-test set (a set of datasets different from the ones used in the meta-training set); (2) the population was converging too fast and, on average, after 20 generations, approximately 70% of the individuals in the population were the same; (3) the system was using an elitist reproduction strategy, and the elitist individual was being found in the very early generations.

Taking into account these three observations, we realized all of them were related to the same problem: overfitting. At first this was not very obvious, since the value of the fitness in the meta-training set is expected to be greater than the values of the fitness in the meta-test set (since the GGP system "saw" the data in the meta-training set many times during evolution). The population could be converging because the fitness and/or the selection procedure were not working properly, and in consequence the long-term survival of an elitist individual found in an early generation would reflect the badly guided search.

We solved (to a large extent) the overfitting problem with a simple and effective solution borrowed from the literature on GP for data mining [1, 3]: at each generation, the data used in the training and validation sets of the datasets in the meta-training set are merged and randomly redistributed into new training and validation sets. This means that, at each generation, the GGP system's individuals are evaluated in a different set of validation data, helping avoid overfitting.

5.5.3 Multiobjective Fitness

As emphasized before, one of the main motivations to use rule induction algorithms to learn knowledge from data is to take advantage of the simplicity and human-readable format of the induced rules. Nevertheless, the fitness function presented in the previous section did not take into account the simplicity of the classification model (rule set or rule list) being built by the evolved rule induction algorithm.

In a second phase of this research, we added to the GGP system a fitness function based on the concept of Pareto optimization [19] (Section 3.3.1), which simultaneously optimizes two objectives:

1. It maximizes the value of the single-objective fitness function based on classification accuracy (see Eq. (5.5)).
2. It minimizes the number of rule conditions in the produced rule set.

This fitness function based on the Pareto optimization concept has to treat one special case: individuals whose rule induction algorithms produce models consisting of only one empty rule, i.e., individuals predicting the class of the most common examples in the training set for all examples in the validation set. For these individuals, the number of conditions in the produced model will always be 0. As we are

Algorithm 5.3: Multiobjective version of the GGP system

```
Let the meta-training set be a set of datasets
pop = N individuals randomly generated from the grammar
for i = 1 to maximum number of generations do
    non-dominated = ∅
    for every individual Ind in pop do
        Run the rule induction algorithm represented by Ind in the meta-training set
        obj1 = average of the normalized accuracy in the meta-training set
        obj2 = average of the number of rule conditions in the meta-training set
    non-dominated = Individuals in pop non-dominated according to the Pareto criterion
                    using obj1 and obj2
    newPop = ∅
    if size(non-dominated) > N/2 then
        Calculate value of f_tb for individuals in non-dominated
        Sort non-dominated according to f_tb
        newPop = N/2 individuals with highest value of f_tb
    else newPop = newPop ∪ non-dominated
    while size(newPop) < N do
        Use tournament selection to select Ind1 and Ind2 from pop
        Apply crossover and mutation operators to Ind1 and Ind2 according to
        user-defined probabilities
        newPop = newPop ∪ Ind1 ∪ Ind2
    pop = newPop
non-dominated = Individuals in pop non-dominated according to the Pareto criterion
                using obj1 and obj2
Calculate value of f_tb for individuals in non-dominated
Sort non-dominated according to f_tb
Return the individual with the highest value of f_tb
```

dealing with a minimization problem, 0 represents the best possible value, and these individuals would always appear in the Pareto front. To avoid this situation, the number of rule conditions of a model with 0 conditions is set to an extremely large value, arbitrarily to 100,000.

Apart from the individuals' evaluation process, some other modifications were added to the single-objective version of the GGP system to deal with the multiobjective approach, as illustrated in the pseudocode of the Multiobjective GGP (MOGGP) system, described in Alg. 5.3. First, in the MOGGP system's selection process, the individuals have to be selected according to a relationship of dominance instead of a simple fitness value. In a tournament selection of size 2, for instance, which is the method used in this system, the winner of the tournament is the individual that dominates the other (recall that an individual Ind_1 dominates an individual Ind_2 if Ind_1 is not worse than Ind_2 in any of the objectives being optimized, and Ind_1 is strictly better than Ind_2 in at least one of the objectives being optimized). In the case that none of the individuals dominates the other, a tiebreaking criterion decides the winner.

This tiebreaking criterion considers the difference in the number of individuals in the entire population that are dominated by an individual and the number of individuals in the population that dominate that individual [20]. This function is named f_{tb}, and it acts like a third (tiebreaking) criterion bering optimized by the GGP system. If neither Ind_1 dominates Ind_2 nor vice versa with respect to accuracy and rule set size, the winner of the tournament is the individual with the largest value of f_{tb}.

The second modification introduced in the MOGGP system concerns both the elitist strategy and the final solution returned to the user. The single-objective version of the GGP system works with elitism, so it preserves the best individual found by the GGP system during the evolutionary process. This individual is the one returned to the user.

In the case of the MOGGP system, at each generation all the solutions in the estimated Pareto front (individuals not dominated by any other individual in the population) are preserved, as long as their number does not exceed half the size of the population. If it does, then the individuals with the highest value of f_{tb} are preserved. If after applying f_{tb} the number of individuals is still higher than half the population size, the individuals with better $fitness$ (see Eq. 5.5) are given priority. At the last generation, this same logic is applied when selecting the best individual to be returned to the user and tested in the meta-test set.

5.6 Crossover and Mutation Operations

In a GGP system, the new individuals produced by the crossover and mutation operators have to be consistent with the grammar. For instance, when performing crossover the system cannot select a subtree with root *EvaluateRule* to be exchanged with a subtree with root *SelectCandidateRules*, because this would create an invalid individual according to the grammar.

Therefore, crossover operations have to exchange subtrees whose roots contain the same nonterminal, apart from *Start*. Crossing over two individuals swapping the subtree rooted at *Start* (actually, the entire tree) would generate exactly the same two individuals, and so it would be useless.

Mutation can be applied to a subtree rooted at a nonterminal or applied to a terminal. In the former case, the subtree undergoing mutation is replaced by a new subtree, produced by keeping the same label in the root of the subtree and then randomly generating the rest of the subtree by a new sequence of applications of production rules, and so producing a new derivation subtree. When mutating terminals, the terminal undergoing mutation is replaced by another "compatible" symbol, i.e., a terminal or nonterminal which represents a valid application of the production rule whose antecedent is that terminal's parent in the derivation tree. The probability of mutating a nonterminal is 90%, while the probability of mutating a terminal is 10%.

However, not all the terminals can be mutated. Terminals like *if*, *then*, *else*, and *while*, which would not introduce any modifications to the new individual, are not considered during mutation operations. The crossover operation is not allowed to exchange subtrees rooted at terminal nodes. However, it is possible that a crossover

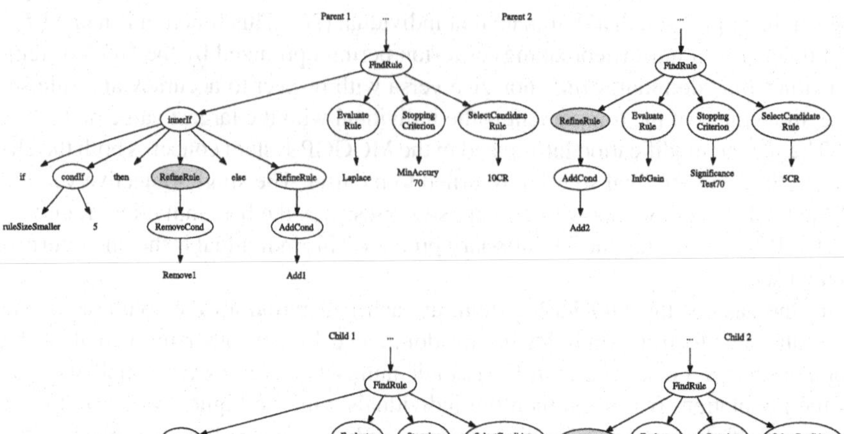

Fig. 5.10 Example of crossover in the proposed GGP system

operation exchanges subtrees rooted at nonterminal nodes that have just one child that is a terminal node. For instance, the nonterminal *EvaluateRule* has just one terminal child node, which could be, say, *accuracy*, *Laplace estimation*, or another evaluation function. In this case, the crossover operation will have the effect of exchanging just the two terminals of the two individuals being crossed over.

Of the 26 nonterminals present in the grammar, eight are more likely to introduce significant changes in the rule induction algorithms represented by the GGP system's individuals. They are, namely, *CreateOneRule*, *InitializeRule*, *innerWhile*, *innerIf*, *FindRule*, *EvaluateRule*, *RefineRule*, and *PrePruneRule*. These nonterminals have a greater chance of being selected to be swapped during crossover operations or replaced during mutation operations. After some preliminary experiments, the probability of crossover or mutation for these nonterminals was set to 70%, the probability of selecting the remaining ones being 30%.

Figure 5.10 shows an example of a crossover operation. Note that just part of the individuals are shown, for the sake of simplicity. The process works as follows. *Parent 1* has a node probabilistically selected for crossover. In the example illustrated, the chosen node is *RefineRule*. The node *RefineRule* is then searched in the derivation tree of *Parent 2*. As *Parent 2* has a node named *RefineRule*, their subtrees are swapped, generating *Child 1* and *Child 2*. If *RefineRule* is not present in the tree of *Parent 2*, a new nonterminal is selected from the tree of *Parent 1*. The GGP system performs at most ten attempts to select a node that can be found in both parents. If after ten attempts it does not happen, both individuals undergo mutation operations.

Recall that there is a set of combinations of terminals and nonterminals of the grammar that are not allowed because they generate invalid individuals (see Section 5.4). These combinations are avoided through a set of constraints considered in the population initialization process. These same constraints are also considered during the crossover and mutation operations.

5.7 Summary

This chapter introduced a Grammar-based Genetic Programming (GGP) system to automatically evolve rule induction algorithms. It described in detail the main components of the system, including the grammar, the individual representation, the population initialization process, the individual evaluation procedure, and the crossover and mutation operators.

The grammar includes components already implemented in many well-known rule induction algorithms, and also elements we thought could work well in the context of rule induction algorithms, but that were not tried before.

An individual encodes a rule induction algorithm, and is represented by a derivation tree created by applying a set of production rules from the grammar. An individual is evaluated using a meta-training set (a set of datasets). For each dataset in the meta-training set, a classification accuracy and a rule set or list is obtained. The GGP system can then use a single objective approach or a multiobjective approach (based on the concept of Pareto optimality) to select the individuals that will be passed to the next generation. The single objective approach aims at maximizing classification accuracy only, while the multiobjective approach aims at both maximizing classification accuracy and minimizing the size of the classification model.

Finally, we showed how the population initialization procedure and the crossover and mutation operations were adapted to generate only individuals that are valid according to the grammar. The next chapter presents the results of experiments performed to evaluate the effectiveness of the proposed GGP system in automatically designing rule induction algorithms and its sensitivity to parameters, and how the effectiveness of GP as a search method can be compared to the effectiveness of a more conventional, simpler search method (hill-climbing search).

References

1. Bhattacharyya, S.: Direct marketing response models using genetic algorithms. In: Proc. of 4th Int. Conf. on Knowledge Discovery and Data Mining (KDD-98), pp. 144–148 (1998)
2. Caruana, R., Niculescu-Mizil, A.: Data mining in metric space: an empirical analysis of supervised learning performance criteria. In: Proc. of the 10th Int. Conf. on Knowledge Discovery and Data Mining (KDD-04), pp. 69–78. ACM Press (2004)
3. Cavaretta, M.J., Chellapilla, K.: Data mining using genetic programming: the implications of parsimony on generalization error. In: P.J. Angeline, Z. Michalewicz, M. Schoenauer, X. Yao,

A. Zalzala (eds.) Proc. of the Congress on Evolutionary Computation (CEC-99), vol. 2, pp. 1330–1337. IEEE Press (1999)

4. Cendrowska, J.: Prism: an algorithm for inducing modular rules. International Journal of Man-Machine Studies **27**, 349–370 (1987)
5. Clark, P., Boswell, R.: Rule induction with CN2: some recent improvements. In: Y. Kodratoff (ed.) Proc. of the European Working Session on Learning on Machine Learning (EWSL-91), pp. 151–163. Springer-Verlag, New York, NY, USA (1991)
6. Clark, P., Niblett, T.: The CN2 induction algorithm. Machine Learning **3**, 261–283 (1989)
7. Cohen, W.W.: Fast effective rule induction. In: A. Prieditis, S. Russell (eds.) Proc. of the 12th Int. Conf. on Machine Learning (ICML-95), pp. 115–123. Morgan Kaufmann, Tahoe City, CA (1995)
8. Fawcett, T.: ROC graphs: notes and practical considerations for data mining researchers. Tech. Rep. HPL-2003-4, HP Labs (2003)
9. Flach, P.: The geometry of ROC space: understanding machine learning metrics through ROC isometrics. In: Proc. 20th Int. Conf. on Machine Learning (ICML-03), pp. 194–201. AAAI Press (2003)
10. Freitas, A.A.: Data Mining and Knowledge Discovery with Evolutionary Algorithms. Springer-Verlag (2002)
11. Fürnkranz, J.: A pathology of bottom-up hill-climbing in inductive rule learning. In: Proc. of the 13th Int. Conf. on Algorithmic Learning Theory (ALT-02), pp. 263–277. Springer-Verlag, London, UK (2002)
12. Goldberg, D.E.: Genetic Algorithms in Search, Optimization, and Machine Learning. Addison-Wesley, Reading, MA (1989)
13. Hekanaho, J.: Background knowledge in GA-based concept learning. In: T. Fogarty, G. Venturini (eds.) 13th Int. Conf. on Machine Learning (ICML-96), pp. 234–242 (1996)
14. Newman, D.J., Hettich, S., Blake, C.L., Merz, C.J.: UCI Repository of machine learning databases. University of California, Irvine, http://www.ics.uci.edu/~mlearn/MLRepository.html (1998)
15. O'Neill, M., Brabazon, A., Ryan, C., Collins, J.: Evolving market index trading rules using grammatical evolution. In: E.J.W. Boers, S. Cagnoni, J. Gottlieb, E. Hart, P.L. Lanzi, G.R. Raidl, R.E. Smith, H. Tijink (eds.) Applications of Evolutionary Computing, *LNCS*, vol. 2037, pp. 343–352. Springer-Verlag (2001)
16. Pappa, G.L.: Automatically evolving rule induction algorithms with grammar-based genetic programming. Ph.D. thesis, Computing Laboratory, University of Kent, Canterbury, UK (2007)
17. Pappa, G.L., Freitas, A.A.: Towards a genetic programming algorithm for automatically evolving rule induction algorithms. In: J. Fürnkranz (ed.) Proc. of the ECML/PKDD-04 Workshop on Advances in Inductive Learning, pp. 93–108. Pisa (2004)
18. Pappa, G.L., Freitas, A.A.: Automatically evolving rule induction algorithms. In: J. Fürnkranz, T. Scheffer, M. Spiliopoulou (eds.) Proc. of the 17th European Conf. on Machine Learning (ECML-06), *Lecture Notes in Computer Science*, vol. 4212, pp. 341–352. Springer Berlin/Heidelberg (2006)
19. Pappa, G.L., Freitas, A.A.: Evolving rule induction algorithms with multi-objective grammar-based genetic programming. Knowledge and Information Systems **19**(3), 283–309 (2009)
20. Pappa, G.L., Freitas, A.A., Kaestner, C.A.A.: Multi-objective algorithms for attribute selection in data mining. In: C.A.C. Coello, G. Lamont (eds.) Applications of Multi-Objective Evolutionary Algorithms, pp. 603–626. World Scientific (2004)
21. Provost, F., Fawcett, T., Kohavi, R.: The case against accuracy estimation for comparing induction algorithms. In: Proc. of the 15th Int. Conf. on Machine Learning (ICML-98), pp. 445–453. Morgan Kaufmann Publishers, San Francisco, CA, USA (1998)
22. Quinlan, J.R.: C4.5: programs for machine learning. Morgan Kaufmann (1993)
23. Tsakonas, A., Dounias, G., Jantzen, J., Axer, H., Bjerregaard, B., von Keyserlingk, D.G.: Evolving rule-based systems in two medical domains using genetic programming. Artificial Intelligence in Medicine **32**(3), 195–216 (2004)

24. Webb, G.I., Brkic, N.: Learning decision lists by prepending inferred rules. In: Proc. of the AI-93 Workshop on Machine Learning and Hybrid Systems, pp. 6–10. World Scientific (1993)
25. Weiss, S.M., Indurkhya, N.: Optimized rule induction. IEEE Expert: Intelligent Systems and Their Applications **8**(6), 61–69 (1993)
26. Whigham, P.A.: Grammatically-based genetic programming. In: J.P. Rosca (ed.) Proc. of the Workshop on Genetic Programming: From Theory to Real-World Applications, pp. 33–41. Tahoe City, California, USA (1995)
27. Witten, I.H., Frank, E.: Data Mining: Practical Machine Learning Tools and Techniques with Java Implementations, 2nd edn. Morgan Kaufmann (2005)
28. Zhang, J.: Selecting typical instances in instance-based learning. In: Proc. of the 9th Int. Workshop on Machine Learning (ML-92), pp. 470–479. Morgan Kaufmann, San Francisco, CA, USA (1992)

Chapter 6
Computational Results on the Automatic Design of Full Rule Induction Algorithms

6.1 Introduction

In Chapter 5, we proposed an evolutionary method for automatically designing rule induction algorithms. The proposed method generates a framework that can be used in at least two different approaches:

1. To design a *robust* rule induction algorithm, which is expected to perform well across multiple datasets (from different application domains)
2. To design a rule induction algorithm tailored to *one specific data set* (from a single application domain)

The Grammar-based Genetic Programming (GGP) system can produce rule induction algorithms following either of these two approaches simply by modifying accordingly the datasets used during the GGP system's training phase. For instance, when creating *robust* rule induction algorithms, a diverse set of datasets is used in the meta-training set required by the algorithm. In theory, the more diverse the datasets, the more robust the produced rule induction algorithm.

In contrast, when creating rule induction algorithms tailored to a specific dataset, only the target dataset is used in the meta-training set. For example, if we want to find a rule induction algorithm tailored to the UCI [16] dataset *promoters*, the data subsets in both the meta-training and the meta-test sets will use data subsets from the dataset *promoters*. Note that, in this context, the use of the terms meta-training and meta-test sets is less justifiable once we have only one dataset in the meta-training set and one dataset in the meta-test set. However, as we are performing some kind of meta-learning anyway, we keep this terminology.

This chapter presents results obtained by the GGP system when following these two different approaches to automatically design rule induction algorithms, and is organized in two main parts. The first part (Section 6.2) reports and analyzes experiments that evaluate the performance of the GGP system when designing robust rule induction algorithms. The second part (Section 6.3) reports and analyzes results obtained by the GGP system designing rule induction algorithms tailored to a specific

G.L. Pappa, A.A. Freitas, *Automating the Design of Data Mining Algorithms*,
Natural Computing Series, DOI 10.1007/978-3-642-02541-9_6,

dataset. Finally, Section 6.4 summarizes the experiments and results described in this chapter.

6.2 Evolving Rule Induction Algorithms Robust Across Different Application Domains

This section reports computational results obtained by the GGP system proposed in Chapter 5 to evolve *robust* rule induction algorithms from multiple datasets. The proposed GGP system was evaluated in five phases. In the first phase, as described in Section 6.2.1, we study the impact of different GGP system parameter values (especially the crossover and mutation rates) in the results obtained by the system in order to select a good parameter setting for further experiments.

In the second phase, as described in Section 6.2.2, we evaluate the rule induction algorithms automatically designed by the GGP system by comparing them with four well-known, manually designed rule induction algorithms: the ordered [5] and unordered [4] versions of CN2, Ripper [6], and C4.5Rules [22].

In the third phase, described in Section 6.2.3, we analyze the rule induction algorithms automatically evolved by the GGP system (denoted GGP-RIs) and compare their structure (in terms of their algorithmic components) with the structure of the human-designed rule induction algorithms.

In the fourth phase, we examine the influence of the number of datasets used in the meta-training set of the GGP system on the structure of the rule induction algorithms produced. Is a GGP system trained with three datasets able to produce algorithms competitive with the ones produced by the GGP system trained with five or seven datasets? This type of question is the subject of Section 6.2.4.

In the fifth phase, we move from changing the GGP system's parameters to changing the GGP system's components. We performed experiments with three variations of the original grammar proposed in Section 5.2, and analyzed how the GGP-RIs produced by these variations compare with the GGP-RIs evolved by the system with the original version of the grammar.

Furthermore, in order to evaluate the effectiveness of the GGP system in designing good rule induction algorithms, Section 6.2.6 reports the results of using a hill-climbing search method to automatically design rule induction algorithms, and compares its results with the rule induction algorithms designed by the GGP system.

In all the results reported in Sections 6.2.1 through 6.2.6, a single-objective fitness function is used by the GGP system. At last, Section 6.2.7 presents results when using a variation of the GGP system evaluated in the previous sections where a multiobjective fitness function is considered.

Table 6.1 Datasets used by the GGP system

Dataset	Examples	Attributes Nomin.	Attributes Numer.	Classes	Def. Acc. (%)
bal-scale	416/209	-	4	3	46
crx	461/229	9	6	2	67.7
glass	145/69	-	9	7	35.2
heart-c	202/101	7	6	2	54.5
hepatitis	104/51	14	6	2	78
ionosphere	234/117	-	34	2	64
lymph	98/50	18	-	4	54
monks-1	124/432	6	-	2	50
monks-2	169/432	6	-	2	67
monks-3	122/432	6	-	2	52
mushroom	5416/2708	23	-	2	52
pima	513/255	-	8	2	65
promoters	70/36	58	-	2	50
segment	1540/770	-	19	7	14.3
sonar	139/69	-	60	2	53
splice	2553/637	63	-	3	52
vehicle	566/280	-	18	4	26
vowel	660/330	3	10	11	9
wisconsin	456/227	9	-	2	65
zoo	71/28	16	-	7	43

6.2.1 Investigating the GGP System's Sensitivity to Parameters

All the experiments performed using the proposed GGP system need two sets of parameters to be defined: (1) the parameters for the GGP system, as defined in Section 3.5, and (2) the datasets used during the training phase of the algorithm. In this section we analyze how the GGP system's parameters – especially the crossover and mutation rates – influence the results obtained by that system.

In order to investigate the influence the GGP system's parameters have in the predictive performance of the rule induction algorithms produced, we first define the datasets that will be used in the GGP system's meta-training and meta-test sets. However, it is not clear how many datasets should be used in each of these meta-sets of data, or what would be the best criteria to distribute the available datasets into these two meta-sets. Intuitively, the larger the number of datasets in the meta-training set, the more robust the evolved rule induction algorithm should be. On the other hand, the smaller the number of datasets in the meta-test set, the less information we have about the ability of the evolved rule induction algorithm to obtain a high predictive accuracy for datasets unseen during the evolutionary run of the GGP system.

Table 6.1 shows the 20 datasets used in the experiments, all obtained from the UCI [16] repository. The figures in the column *Examples* indicate the number of examples present in the training and validation datasets (numbers before and after the "/", respectively), followed by the number of nominal attributes, numerical attributes, and classes. The last column shows the default accuracy (the accuracy

Table 6.2 Predictive accuracy rates (%) obtained by the rule induction algorithms evolved by the GGP system in the meta-test set, for different values of crossover and mutation rates

Dataset	Cross. 0.5 Mut. 0.45	Cross. 0.6 Mut. 0.35	Cross. 0.7 Mut. 0.25	Cross. 0.8 Mut. 0.15	Cross. 0.9 Mut. 0.05
crx	77.75±3.81	80.95±0.82	77.46±3.8	79.54±1.62	81.85±0.97
heart-c	74.67±1.7	75.47±1.77	76.72±1.5	75.44±0.66	77.37±1.19
ionosphere	87.2±2.13	86.59±1.97	87.04±2.2	85.88±1.64	87.61±2.24
monks-1	99.82±0.18	99.82±0.18	99.93±0.07	99.64±0.36	100±0
mushroom	99.23±0.05	99.96±0.01	99.99±0	100±0	99.75±0.05
promoters	80.5±2.29	83.59±2.7	78.98±2.93	78.4±1.9	75.97±1.82
segment	93.98±0.57	95.04±0.48	95.06±0.26	95.34±0.31	93.95±0.37
sonar	72.83±1.91	72.52±1.87	72.34±1.91	71.48±2.3	69.24±1.14
splice	89.16±0.47	89.82±0.42	88.68±0.31	89.82±0.44	82.19±0.47
wisconsin	95.11±0.47	95.97±0.38	95.58±0.74	95.02±0.5	95.23±0.7

obtained when using the most frequent class in the training set to classify new examples in the validation or test set). It is important to note that during the evolution of the rule induction algorithm by the GGP system, for each dataset in the meta-training set, each candidate rule induction algorithm (i.e., each GGP individual) is trained with 70% of the examples, and then validated with the remaining 30% – with the exception of the datasets with predefined training and validation sets, such as *monks*. In this case the original sets were kept. Recall that, during a run of the GGP system, the set of examples used to train a rule induction algorithm is called the training set, while the set of examples used to validate the classification model built by a rule induction algorithm is called the validation set. These two sets of data vary from generation to generation to avoid overfitting of the rule induction algorithm being evolved to the data (see Section 5.5.2). In contrast, in the meta-test set, the evolved rule induction algorithms are evaluated using a well-known five-fold cross validation procedure [25].

In total, ten datasets were used to train and ten to test the GGP system. We selected the datasets that compose the meta-training set based on the execution time of rule induction algorithms, so that we included in the meta-training set the datasets leading to faster runs of the rule induction algorithms. The datasets selected to be part of the training set were *balance-scale*, *glass*, *hepatitis*, *lymph*, *monks-2*, *monks-3*, *pima*, *vowel*, *vehicle*, and *zoo*. The meta-test set was composed of *crx*, *heart-c*, *ionosphere*, *monks-1*, *mushroom*, *promoters*, *segment*, *sonar*, *splice*, and *wisconsin*.

After creating the meta-training and meta-test sets, we turned to the GGP system's parameters: population size, number of generations, tournament size, and crossover, mutation, and reproduction rates. In all the experiments reported in this section, the population size is set to 100, the number of generations to 30, and the tournament size to 2. These three figures were chosen when evaluating the GGP system's evolution in preliminary experiments, but are not optimized. Regarding crossover, mutation, and reproduction rates, GP algorithms usually use a high rate of crossover and low rates of mutation and reproduction. However, the balance between these three numbers is an open question, and may be very problem-dependent [1].

The experiments shown in this section aimed to find a good trade-off between the crossover and mutation rates. In order to do so, the reproduction rate was set to 0.05, and the balance between the crossover and mutation rates varied. Table 6.2 shows the average predictive accuracy in the meta-test set of the rule induction algorithms evolved by the GGP system with different values for crossover and mutation rates. For each cell of Table 6.2, the reported accuracy is the average over five runs of the GGP system, using a different random seed to initialize the population at each run. The numbers after the symbol $\pm$ are standard deviations. The header of each column shows the crossover and mutation rates used by the GGP system. Recall that for each dataset in the meta-test set a five-fold cross-validation procedure is performed to evaluate the rule induction algorithms produced by the GGP system. Note that the results obtained for the datasets in the meta-training set are omitted here. This is because, as the training data was accessed many times during evolution, these results would be over-optimistic and would not be a fair measure of generalization performance.

All the results in this section are compared using a statistical hypothesis test, namely the well-known paired two-tailed Student's t-test [25]. In the analysis of the results, for the sake of simplicity, we will refer to the GGP system using a crossover rate of 0.5 as GGP-0.5, to the GGP system using a crossover rate of 0.6 as GGP-0.6, and so on. The corresponding mutation rate is given by subtracting the crossover rate from 0.95; recall that the reproduction rate is 0.05.

Applying the t-test to the results in Table 6.2, we find significant differences in the predictive accuracies associated with different parameter configurations of the GGP system in three datasets: *mushroom*, *segment*, and *splice*. In *mushroom*, the accuracies of all the other four parameter configurations are significantly better than the accuracy of GGP-0.5 with significance level 0.01 (i.e., the t-test produced a p value smaller than 0.01), and the accuracies of GGP-0.6, GGP-0.7, and GGP-0.8 are also significantly better than the one of GGP-0.9, with significance level 0.05. In *segment*, GGP-0.8's accuracy is significantly better than GGP-0.5's accuracy, while in *splice* the accuracies in all other parameter configurations are significantly better than the one obtained by GGP-0.9 with significance level 0.01.

Based on these results, we cannot say that any of the GGP parameter configurations evaluated is better than the others. For some datasets, one of them can be slightly better than the others, but overall the results obtained were very similar. From these observations we can conclude that the proposed GGP system is very robust to the setting of crossover and mutation parameters when trained with the previously mentioned datasets in the meta-training set.

6.2.2 Comparing GGP-Designed Rule Induction Algorithms with Human-Designed Rule Induction Algorithms

In Section 6.2.1 we compared the results obtained by the rule induction algorithms evolved by the GGP system with different parameter configurations. These comparisons were useful to show the robustness of the system with respect to variations in the crossover and mutation rates, but they do not give any insights about how competitive the automatically designed algorithms are when compared to some well-known human-designed rule induction algorithms. In this section, we compare the results obtained by the GGP-designed rule induction algorithms (GGP-RIs) with four well-known rule induction algorithms: the ordered [5] and unordered [4] versions of CN2, Ripper, and C4.5Rules.

From these four algorithms, C4.5Rules is the only one that does not follow the sequential covering approach, which is the approach followed by the GGP-RIs. However, as C4.5Rules has been used as a benchmark algorithm for classification problems for many years, we also included it in our set of baseline comparison algorithms.

It is also important to observe that the current version of the grammar does not include all the components present in Ripper, but does include all the components present in both versions of CN2. In other words, the space of candidate rule induction algorithms searched by the GGP system includes CN2, but it does not include C4.5Rules or the complete version of Ripper.

Table 6.3 reports the number of times the GGP-RIs produced with specific crossover and mutation rates are statistically significantly better (worse) than the baseline algorithms using the paired two-tailed Student's t-test with significance level 0.01. The better (worse) results are obtained out of 40 comparisons: 10 datasets times 4 baseline classification algorithms. The cases that do not appear in Table 6.3 are the ones where the results obtained by the GGP-RIs were statistically not significantly different from the ones obtained by the baseline algorithms. As expected, the numbers of significantly better (or worse) results do not vary a lot from one parameter configuration to another (as GGP systems with different parameter configurations produce similar results).

Observe that all the GGP-RIs but one obtained predictive accuracies statistically significantly better than the baseline algorithms in three cases. The exception is the rule induction algorithms produced by GGP-0.9, which obtained better results in two cases instead of three. However, there was a greater variation among the number of results obtained by the GGP-RIs that were significantly worse than the baseline algorithms, varying from one to five. According to their number of losses, GGPs using 0.7 and 0.8 crossover rates produced GGP-RIs slightly better than the others, with three results statistically better and one result statistically worse than the ones obtained by the baseline algorithms.

Instead of showing a more detailed comparison of the GGP-RIs with the baseline algorithms for all the parameter configurations, here we choose to present the detailed results of GGP-0.7 only, since the results produced by the other pa-

Table 6.3 Comparing the GGP-RIs trained with different parameter configurations to the baseline algorithms using Student's t-test

GGP-RIs	
Cross/Mut	Better(worse)
0.5/0.45	3(4)
0.6/0.35	3(3)
0.7/0.25	3(1)
0.8/0.15	3(1)
0.9/0.05	2(5)

Table 6.4 Comparing predictive accuracy rates (%) for the datasets in the meta-test set: results obtained with crossover rate = 0.7 and mutation rate = 0.25

Dataset	GGP-RIs	OrdCN2	UnordCN2	Ripper	C45Rules
crx	77.46±3.8	80.16 ± 1.27	80.6 ± 0.93	84.37 ± 1.21	84.82 ± 1.53
heart-c	76.72±1.5	77.9 ± 1.96	77.54 ± 2.85	77.53 ± 1.1	74.2 ± 5.43
ionosphere	87.04±2.2	87.6 ± 2.76	90.52 ± 2.03	89.61 ± 1.75	89.06 ± 2.71
monks-1	99.93±0.07	100 ± 0	100 ± 0	93.84 ± 2.93	100 ± 0
mushroom	99.98±0.02	100 ± 0	100 ± 0	99.96 ± 0.04	98.8 ± 0.06
promoters	78.98±2.93	81.9 ± 4.65	74.72 ± 4.86	78.18 ± 3.62	83.74 ± 3.46
segment	95.06±0.26	95.38 ± 0.28	85.26 ± 0.87	95.44 ± 0.32	88.16 ± 7.72
sonar	72.34±1.91	70.42 ± 2.66	72.42 ± 1.4	72.88 ± 4.83	72.4 ± 2.68
splice	88.68±0.31	90.32 ± 0.74	74.82 ± 2.94	93.88 ± 0.41	89.66 ± 0.78
wisconsin	95.58±0.74	94.58 ± 0.68	94.16 ± 0.93	93.99 ± 0.63	95.9 ± 0.56

rameter configurations are similar. This parameter configuration, which uses a crossover rate of 0.7 and a mutation rate of 0.25, will also be used in further experiments.

Table 6.4 shows the average predictive accuracy obtained by the rule induction algorithms produced by the GGP system (GGP-RIs) in five different runs of the system (with a different random seed in each run), followed by the results of runs of ordered and unordered CN2, Ripper, and C4.5Rules (using default parameter values in all these algorithms). Cells in dark gray represent results in which the GGP-0.7 obtained significantly better results than the human-designed baseline classification algorithms according to a paired two-tailed Student's t-test with 0.01 significance level. In turn, cells in light gray represent a statistically significantly better result of a baseline algorithm over the GGP-RIs.

We observe that the GGP-RIs obtain significantly better accuracies than unordered CN2 in *segment* and *splice* and C4.5Rules in *mushroom*. Naturally, these three cases involve baseline algorithms with the worst accuracy for the respective dataset. The single result significantly worse than the results obtained by one of the baseline algorithms is obtained by the GGP-RIs in *splice*, which is significantly worse than the Ripper result for that dataset. However, we can observe that the accuracies obtained by Ripper in *splice* are much superior to the ones found by all the other baseline algorithms, and as the GGP-RIs did not have all the algorithmic components present in Ripper in its search space, it was not possible to produce a GGP-RI with such a high accuracy in *splice*.

This experiment leads us to conclude that the GGP-RIs can easily outperform classifiers which are not competitive with the other baseline algorithms. For example, in *splice* the predictive accuracy of unordered CN2 is 74.82% ± 2.94, while the other algorithms obtain accuracies close to 90%. In this case, the GGP-RIs can easily find a better accuracy than the one found by unordered CN2.

Finally, recall that the search space of the GGP system includes both Unordered and Ordered CN2. Hence, it seems fair to expect that the GGP-RIs should not obtain a predictive accuracy significantly worse than either version of CN2. Indeed, this was the case in the experiments reported in this section. Considering the results of Table 6.4, the GGP-RIs significantly outperformed unordered CN2 in two cases (dark gray cells in Table 6.4), and there was no case where either version of CN2 significantly outperformed the GGP-RIs.

6.2.3 To What Extent Are GGP-RIs Different from Manually Designed Rule Induction Algorithms?

The main goal of this work was to automatically design rule induction algorithms robust enough to obtain a competitive predictive accuracy in new datasets when compared to manually designed rule induction algorithms. Section 6.2.2 showed that it is possible to automatically generate competitive rule induction algorithms. However, how similar are they to well-known, manually designed algorithms? Are the GGP-RIs innovations present in any aspect of manually designed algorithms? This section describes the way the CN2 and Ripper algorithms induce rules, and then compares these algorithms with the discovered GGP-RIs. C4.5Rules will not be part of this comparison, as it is does not follow the sequential covering approach. Rather, C4.5Rules extracts a rule set from a decision tree built from the data [22].

Algorithm 6.1 shows the pseudocode for the ordered CN2 algorithm [4]. Note that this algorithm, as all the others described in this section, are instantiations of Alg. 2.4 (in the case of a rule list) or Alg. 2.2 (in the case of a rule set), described in Chapter 2.

Algorithm 6.1 starts to produce rules with an empty condition, adds one condition at a time to it, evaluates the rule with the new condition using the Laplace estimation, and selects the best five produced rules to go into the refinement process. Notice that a rule is considered a candidate to be the best rule only if it is significant according to a statistical significance test. However, the default setting of CN2 completely ignores this test. The size of the beam shown in the pseudocode is 5 because this value is used as the CN2 default one. The algorithm keeps inserting new rules into the rule list while there are uncovered examples in the training set.

The unordered version of CN2 presents just one major modification when compared to Alg. 6.1. An outer loop *for* is inserted to repeat the algorithm described in Alg. 6.1 as many times as the number of classes presented in the data. It is impor-

Algorithm 6.1: Main part of the pseudocode of the ordered CN2 algorithm

```
RuleList = ∅
repeat
    bestRule = an empty rule
    candidateRules = ∅
    candidateRules = candidateRules ∪ bestRule
    while candidateRules ≠ ∅ do
        for each candidateRule CR do
            newCandidateRules = ∅
            Add 1 condition at a time to CR
            Evaluate CR using the Laplace estimation
            if CR is significant then
                if CR better than bestRule then
                    bestRule = CR
                newCandidateRules = newCandidateRules ∪ CR
        candidateRules = 5 best rules selected from newCandidateRules
    RuleList = RuleList ∪ bestRule
    Remove from the training set all examples covered by bestRule
until all examples in the training set are covered
```

tant to point out that, when applying the rule set generated by unordered CN2 to unseen examples in the test set, in cases where more than one rule predicting different classes cover the same example, the class of the example is decided using the following method. For all the rules that cover the example, the number of covered examples in each class is counted. Then, for each class in turn, the number of examples covered by each of the fired rules is summed up. The class which presents the largest number of examples covered by those rules is chosen as the new example's class.

As shown in Alg. 6.1, CN2 is a fairly simple algorithm. Algorithm 6.2 shows the main part of the pseudocode of Ripper [6]. As observed, this is a much more sophisticated algorithm. It actually builds a rule list of ordered classes, that is, for each class in turn it builds a rule list that separates the current class from the other classes in the dataset. It works in three phases: it first grows a rule which covers no negative examples by adding one condition at a time to it, and evaluates the possible candidate rules using the information gain criterion. Once a rule that covers no negative examples is found, it is pre-pruned by removing from it a set of final conditions. During the pruning phase, new rules are evaluated using the formula $(p-n)/(p+n)$, where p is the number of positive examples covered by the pruned rule and n the number of negative examples covered by the pruned rule. Rules are inserted into the rule set until the minimum description length (MDL) of the rule set (described in Section 2.4.3) is larger than a parameter d (representing a number of bits) plus the size of the smallest MDL found so far, or there are no positive examples uncovered. Once the rule set for one class is complete, it goes through an optimization phase. During this phase, each rule is analyzed in turn, and can be replaced by another rule or revised [6]. This decision depends again on the MDL heuristic, which can also lead to a rule being removed from the model.

Algorithm 6.2: Main part of the pseudocode of the Ripper algorithm

```
Let m be the number of classes
Let p and n be the number of positive and negative examples covered by a rule
Sort classes in ascendent order, C_1,...,C_m, according to their number of examples
RuleSet = ∅
for i = 1 to m − 1 do
    Positive = examples from C_i
    Negative = examples from remaining classes
    RuleSet' = ∅
    repeat
        Divide the training data in Grow and Prune
        R = an empty rule
        while R covers negative examples do
            newCandidateRules = ∅
            newCandidateRules = Add 1 condition at a time to R
            Evaluate newCandidateRules using information gain in Grow
            R = best rule in newCandidateRules
        R' = Rule produced when removing last condition from R
        Define pruneEval(x) of a rule x as (p − n)/(p + n)
        while pruneEval(R') > pruneEval(R) in Prune do
            R = R'
            R'= Rule produced when removing last condition from R
        RuleSet' = RuleSet' ∪ R
        Remove examples covered by R from training set
    until Positive ≠ ∅ OR MDL of RuleSet' is d bits > the smallest MDL found so far
    RuleSet = RuleSet ∪ Optimized RuleSet'
    Remove all examples covered by RuleSet' from the training set
Make class C_m the default class
```

It is important to emphasize that the current version of the grammar does not use the MDL criterion to stop producing rules, and also does not implement all the steps required by Ripper's optimization process. Implementing the MDL heuristic and the complete Ripper optimization process are extensions left for future research.

Here we describe one of the GGP-RIs produced by the experiments reported in the previous section. Algorithm 6.3 creates an empty rule, refines it according to the number of conditions present in the rule being refined, and evaluates it using the Laplace estimation. It selects the best two rules to undergo refinements, and requires that the accuracy of the selected rule be greater than 80%. Rules are generated until all the examples in the training set are covered. The main difference between Alg. 6.3 and our baseline comparison algorithms is in the way the rules are refined. In Alg. 6.3, an *if* condition determines whether to add one or two conditions at a time to a candidate rule according to the number of conditions present in the rule being refined.

It should be noted that the strategy of switching the number of conditions added (in a single step) to the current rule, depending on the number of conditions present in the rule being refined, is *innovative*. In general, manually designed rule induction algorithms do not have this flexibility; they simply use a fixed number of conditions to be added to the current rule throughout the run of the algorithm, regardless of the

Algorithm 6.3: Example of a decision list algorithm created by the GGP system

```
RuleList = ∅
repeat
    bestRule = an empty rule
    candidateRules = ∅
    candidateRules = candidateRules ∪ bestRule
    while candidateRules ≠∅ do
        for each candidateRule CR do
            newCandidateRules = ∅
            if size(CR) < 2 then
                Add 1 condition at a time to CR
            else
                Add 2 conditions at a time to CR
            Evaluate CR using the Laplace estimation
            if CR better than bestRule then
                bestRule = CR
            newCandidateRules = newCandidateRules ∪ CR
        candidateRules = 2 best rules selected from newCandidateRules
    if accuracy(bestRule) < 0.8 then break
    else RuleList = RuleList ∪ bestRule
    Remove from the training set all examples covered by bestRule
until all examples in the training set are covered
```

current rule size. The flexible strategy discovered by the GGP system first creates a rule by adding one condition at a time to it. However, when the number of conditions in the rule reaches two, the GGP system changes strategy and starts adding two conditions at a time to it. Although adding two conditions at a time is less greedy (and potentially more effective) than adding one condition at a time, the strategy used above is counterintuitive. This is because, as the number of conditions in the rule increases, the statistical support of the rule (the number of examples satisfying its antecedent) is reduced. As a result, it is intuitively easier to find one attribute-value pair with good predictive power to be added to the current rule than a combination of two attribute-values.

Table 6.5 shows the predictive accuracies obtained by Alg. 6.3 in the datasets of the meta-test set using a five-fold cross-validation procedure. Although the strategy used by the algorithm is counterintuitive, when comparing it with the baseline (manually designed) algorithms, it roughly reproduces the results presented in Table 6.4. It has significantly better accuracies than the weakest algorithms, obtaining a significantly better accuracy in three cases, and significantly worse accuracy than Ripper in *splice*. Besides, it also obtains accuracies significantly better than Ripper in *monks-1* and all algorithms but C4.5Rules in *wisconsin*. At the same time, it obtains significantly worse accuracies than Ripper and C4.5Rules in *crx*. In summary, it presents statistically better accuracies in seven out of 40 comparisons, and statistically worse accuracies in three out of 40 comparisons.

Besides the algorithm described in Alg. 6.3, some other evolved algorithms presented innovate behavior in the way they produce or prune rules. However, in gen-

Table 6.5 Predictive accuracy rates (%) produced by Alg. 6.3 and human-designed algorithms

Dataset	GGP-RIs	OrdCN2	UnordCN2	Ripper	C45Rules
crx	79.44±2.24	80.16 ± 1.27	80.6 ± 0.93	84.37 ± 1.21	84.82 ± 1.53
balance-scale	79.14±4.93	82.08 ± 1.38	79.48 ± 1.7	81.12 ± 0.88	78.84 ± 1.7
heart-c	76.6±8.01	77.9 ± 1.96	77.54 ± 2.85	77.53 ± 1.1	74.2 ± 5.43
ionosphere	86.94±5.41	87.6 ± 2.76	90.52 ± 2.03	89.61 ± 1.75	89.06 ± 2.71
monks-1	99.64±0.8	100 ± 0	100 ± 0	93.84 ± 2.93	100 ± 0
mushroom	100±0	100 ± 0	100 ± 0	99.96 ± 0.04	98.8 ± 0.06
promoters	83.92±7.34	81.9 ± 4.65	74.72 ± 4.86	78.18 ± 3.62	83.74 ± 3.46
segment	94.54±1.23	95.38 ± 0.28	85.26 ± 0.87	95.44 ± 0.32	88.16 ± 7.72
sonar	73.12±7.94	70.42 ± 2.66	72.42 ± 1.4	72.88 ± 4.83	72.4 ± 2.68
splice	90.1±1.08	90.32 ± 0.74	74.82 ± 2.94	93.88 ± 0.41	89.66 ± 0.78
wisconsin	96.2±1	94.58 ± 0.68	94.16 ± 0.93	93.99 ± 0.63	95.9 ± 0.56

eral, we could say that most of the algorithms produced were very similar to CN2. But why did this happen?

The UCI datasets [16] are very popular in the machine learning community, and they have been used to benchmark classification algorithms for a long time. To a certain extent, most of the rule induction algorithms were first designed or later modified targeting these datasets. The fact that the evolved rule induction algorithms are so similar to CN2, for instance, is evidence that CN2 is actually a very good algorithm in terms of average predictive accuracy in a set of datasets available in the UCI repository. At the same time, as the rule induction algorithms produced by the GGP system have shown, there are many other variations of the basic sequential covering pseudocode that obtain accuracies competitive with the ones produced by CN2, Ripper, or C4.5Rules. In any case, the GGP system has been more successful in generating more innovative rule induction algorithms in other experiments, as will be mentioned in Section 6.3.1.

In general, the evolved algorithms shown in Alg. 6.3 did not obtain significantly better accuracies than the baseline (manually designed) rule induction algorithms, but the former obtained slightly better results than the latter, overall. This can be observed in Table 6.4, which overall contains more significant wins (dark gray cells) than significant losses (light gray cells) for the evolved algorithms.

6.2.4 Meta-training Set Variations

In Section 6.2.1 we explained that the GGP system needed two sets of elements in order to automatically evolve rule induction algorithms: the conventional GP algorithm's parameters – such as population size, number of generations and crossover and mutation rates, and the number of datasets (and the datasets themselves) used in the meta-training set of the GGP system.

As the datasets in the meta-training set are used to calculate the fitness of a GGP system's individual (candidate rule induction algorithm), they are a key point in the GGP system's evolutionary process. Changing the datasets in the meta-training

Table 6.6 Comparing predictive accuracy rates (%) in the meta-test set when training the GGP with six datasets

Dataset	GGP-RIs	OrdCN2	UnordCN2	Ripper	C45Rules
crx	81.07±0.9	80.16 ± 1.27	80.6 ± 0.93	84.37 ± 1.21	84.82 ± 1.53
heart-c	77.18±0.9	77.9 ± 1.96	77.54 ± 2.85	77.53 ± 1.1	74.2 ± 5.43
hepatitis	82.51±1.89	81.94 ± 5.02	83.34 ± 1.83	86.03 ± 1.14	83.36 ± 0.9
ionosphere	87.33±2.17	87.6 ± 2.76	90.52 ± 2.03	89.61 ± 1.75	89.06 ± 2.71
monks-1	100±0	100 ± 0	100 ± 0	93.84 ± 2.93	100 ± 0
monks-3	96.89±0.78	97.46 ± 0.74	99.1 ± 0.4	98.54 ± 0.46	94 ± 4.89
mushroom	100±0	100 ± 0	100 ± 0	99.96 ± 0.04	98.8 ± 0.06
pima	63.86±3.16	69.34 ± 2.13	74.6 ± 0.38	73.91 ± 1.65	71.04 ± 1.67
promoters	81.61±3.08	81.9 ± 4.65	74.72 ± 4.86	78.18 ± 3.62	83.74 ± 3.46
segment	95.14±0.39	95.38 ± 0.28	85.26 ± 0.87	95.44 ± 0.32	88.16 ± 7.72
sonar	73.12±1.44	70.42 ± 2.66	72.42 ± 1.4	72.88 ± 4.83	72.4 ± 2.68
splice	88.76±0.38	90.32 ± 0.74	74.82 ± 2.94	93.88 ± 0.41	89.66 ± 0.78
wisconsin	95.53±0.74	94.58 ± 0.68	94.16 ± 0.93	93.99 ± 0.63	95.9 ± 0.56
zoo	92.89±0.87	92.64 ± 1.33	92.52 ± 2.21	89.47 ± 1.66	92.56 ± 1.45

Table 6.7 Comparing predictive accuracy rates (%) in the meta-test set when training the GGP system with eight datasets

Dataset	GGP-RIs	OrdCN2	UnordCN2	Ripper	C45Rules
crx	80.57±0.9	80.16 ± 1.27	80.6 ± 0.93	84.37 ± 1.21	84.82 ± 1.53
heart-c	76.38±0.82	77.9 ± 1.96	77.54 ± 2.85	77.53 ± 1.1	74.2 ± 5.43
ionosphere	86.75±2.12	87.6 ± 2.76	90.52 ± 2.03	89.61 ± 1.75	89.06 ± 2.71
monks-1	99.82±0.18	100 ± 0	100 ± 0	93.84 ± 2.93	100 ± 0
mushroom	100±0	100 ± 0	100 ± 0	99.96 ± 0.04	98.8 ± 0.06
pima	68.6±0.94	69.34 ± 2.13	74.6 ± 0.38	73.91 ± 1.65	71.04 ± 1.67
promoters	80.23±3.29	81.9 ± 4.65	74.72 ± 4.86	78.18 ± 3.62	83.74 ± 3.46
segment	94.99±0.41	95.38 ± 0.28	85.26 ± 0.87	95.44 ± 0.32	88.16 ± 7.72
sonar	72.73±1.74	70.42 ± 2.66	72.42 ± 1.4	72.88 ± 4.83	72.4 ± 2.68
splice	89.04±0.48	90.32 ± 0.74	74.82 ± 2.94	93.88 ± 0.41	89.66 ± 0.78
wisconsin	95.76±0.75	94.58 ± 0.68	94.16 ± 0.93	93.99 ± 0.63	95.9 ± 0.56
zoo	92.65±1.12	92.64 ± 1.33	92.52 ± 2.21	89.47 ± 1.66	92.56 ± 1.45

set can significantly change the values of the fitness of the GGP individuals and, consequently, the final rule induction algorithm produced by the GGP system.

Hence, in this section we study the impact of changing the number of datasets in the meta-training set. We vary the number of datasets in the meta-training set to 6, 8, 12, and 14 (making the number of datasets in the meta-test set 14, 12, 8, and 6, respectively). The choice of which datasets would be removed or added to the original meta-training set was, as before, based on execution time. Note that all the GGP system's parameter values were fixed in all the experiments reported in this section, and the crossover and mutation rates used here are the same ones chosen when presenting the comparisons among the GGP-RIs and other baseline algorithms in Section 6.2.2, i.e., crossover and mutation rates of 0.7 and 0.25, respectively.

Tables 6.6, 6.7, 6.8, and 6.9 present the results using 6 (14), 8 (12), 12 (8) and 14 (6) datasets in the meta-training (meta-test) set, respectively. As before, cells in dark

Table 6.8 Comparing predictive accuracy rates (%) in the meta-test set when training the GGP system with 12 datasets

Dataset	GGP-RIs	OrdCN2	UnordCN2	Ripper	C45Rules
crx	78.12±3.37	80.16 ± 1.27	80.6 ± 0.93	84.37 ± 1.21	84.82 ± 1.53
heart-c	75.08±2.44	77.9 ± 1.96	77.54 ± 2.85	77.53 ± 1.1	74.2 ± 5.43
mushroom	100±0	100 ± 0	100 ± 0	99.96 ± 0.04	98.8 ± 0.06
promoters	81.05±2.94	81.9 ± 4.65	74.72 ± 4.86	78.18 ± 3.62	83.74 ± 3.46
segment	95.11±0.43	95.38 ± 0.28	85.26 ± 0.87	95.44 ± 0.32	88.16 ± 7.72
sonar	73.69±2.4	70.42 ± 2.66	72.42 ± 1.4	72.88 ± 4.83	72.4 ± 2.68
splice	89.3±0.39	90.32 ± 0.74	74.82 ± 2.94	93.88 ± 0.41	89.66 ± 0.78
wisconsin	95.9±0.6	94.58 ± 0.68	94.16 ± 0.93	93.99 ± 0.63	95.9 ± 0.56

gray represent results in which the GGP-RIs obtained significantly better accuracies than the other baseline algorithms according to a paired two-tailed Student's t-test with 0.05 significance level. In turn, cells in light gray represent a significantly better accuracy of a baseline rule induction algorithm over the GGP-RIs.

Analyzing these results and comparing them to the ones presented in Table 6.4 (results of the GGP system trained with ten datasets), all the results are consistent: the cases in which the GGP-RIs obtained significantly better or worse accuracies than the baseline algorithms in particular datasets remain the same. However, when *pima* is inserted in the meta-test set (in experiments using six and eight datasets in the meta-training set), the GGP-RIs obtained significantly worse accuracies than unordered CN2 and Ripper, as observed in Tables 6.6 and 6.7. Looking at the ten GGP-RIs produced when using six and eight datasets in the meta-training set, respectively, all of them produced rule lists. It seems that, in the case of *pima*, rule sets are likely to perform better than rule lists, and that is why the GGP-RIs are competitive with ordered CN2 but not with unordered CN2.

In Tables 6.6, 6.7, and 6.9 the GGP-RIs also obtained accuracies significantly worse than the ones obtained by Ripper in *crx*. The same is true for the accuracies of the GGP-RIs presented in Table 6.7 when running C4.5Rules in *crx*.

To summarize, when the meta-training set includes six and eight datasets (Tables 6.6 and 6.7, respectively), we can observe, overall, some drop in the accuracies obtained by the GGP-RIs in comparison with the accuracies obtained by the baseline algorithms. In particular, in both Tables 6.6 and 6.7, the GGP-RIs significantly outperform the baseline algorithms in just three cases, and the reverse is true in five cases – overall a slightly negative result which can be summarized by the score -2. In contrast, when the meta-training set includes 12 and 14 datasets (Tables 6.8 and 6.9, respectively), the GGP-RIs obtain the slightly positive scores of $+2$ and $+1$, respectively.

Hence, overall we can conclude that the results produced in this section are not significantly different from the ones produced when using ten datasets in the meta-training and ten datasets in the meta-test set. Our intuition was that decreasing the number of datasets in the meta-training set might lead the GGP system to create algorithms with poorer performance in the meta-test set, but that was not the case in general, with a few exceptions. The results obtained in experiments with fewer

Table 6.9 Comparing predictive accuracy rates (%) in the meta-test set when training the GGP system with 14 datasets

Dataset	GGP-RIs	OrdCN2	UnordCN2	Ripper	C45Rules
crx	80.81±1.14	80.16 ± 1.27	80.6 ± 0.93	84.37 ± 1.21	84.82 ± 1.53
mushroom	100±0	100 ± 0	100 ± 0	99.96 ± 0.04	98.8 ± 0.06
promoters	81.72±3.3	81.9 ± 4.65	74.72 ± 4.86	78.18 ± 3.62	83.74 ± 3.46
segment	95.21±0.43	95.38 ± 0.28	85.26 ± 0.87	95.44 ± 0.32	88.16 ± 7.72
sonar	71.94±1.46	70.42 ± 2.66	72.42 ± 1.4	72.88 ± 4.83	72.4 ± 2.68
splice	89.33±0.38	90.32 ± 0.74	74.82 ± 2.94	93.88 ± 0.41	89.66 ± 0.78

datasets in the meta-training set are in general almost as good as the ones obtained with more datasets in the meta-training set.

A possible explanation for this fact is that the rule induction algorithms produced with fewer datasets in the meta-training set are simpler than the ones built with a larger number of datasets in the meta-training set. Simpler algorithms seem more likely to be more robust and to have a better performance with very different kinds of data.

6.2.5 GGP System's Grammar Variations

Sections 6.2.1 and 6.2.4 showed the proposed GGP system is not too sensitive to variations in the values of its parameters. In this section we show how the GGP system responds to a change in one of its main components: the grammar. The grammar determines the GGP system's search space. Hence, changing the grammar might completely change the rule induction algorithms produced by the system.

It is not easy to judge whether a particular version of the grammar is better or worse than another. There are two main reasons for that. First, the GGP system can adapt according to the functionalities present in the grammar. As will be shown later, turning off the GGP system's ability to produce rule induction algorithms following the top-down approach led the GGP system to produce completely different but still overall competitive rule induction algorithms. Second, the grammar has a set of core components that form the basis for the construction of rule induction algorithms. By making use of these components, the system will be able to, most of the time, produce a relatively simple but competitive rule induction algorithm. For these reasons, we will compare the GGP-RIs generated by the system when using different versions of the grammar by considering the predictive accuracy they obtained in the datasets in the meta-test set.

Three types of experiments were executed to evaluate the impact of adding or removing certain production rules to or from the grammar. In the first experiment, we removed from the grammar all its "new components," described in Section 5.2.1. These "new components" correspond to functionalities that were added to the grammar but not used before (to the best of our knowledge) by the manually designed

Table 6.10 Comparing predictive accuracy rates (%) of the GGP system in the meta-test set, with different versions of the grammar

Dataset	Grammar			
	Original	Basic	NoPrune	BottomUp
crx	77.46±3.8	80.19±1.11	80.14±0.73	81.33±1.14
heart-c	76.72±1.5	76.44±1.53	77.37±1.39	75.3±0.96
ionosphere	87.04±2.2	85.72±1.81	86.06±2.18	84.85±1.6
monks-1	99.93±0.07	100±0	100±0	100±0
mushroom	99.98±0.02	100±0	99.9±0.02	99.79±0.1
promoters	78.98±2.93	80.16±1.22	74.71±0.76	52.11±1.96
segment	95.06±0.26	95.95±0.19	94.32±0.24	88.47±1.2
sonar	72.34±1.91	76.38±3.04	74.45±2.64	60.86±0.85
splice	88.68±0.31	90.08±0.44	82.85±0.44	50.72±0.49
wisconsin	95.58±0.74	94.61±0.51	94.14±0.44	91.18±0.55

rule induction algorithms. They include the symbols *innerIf*, *MakeFirstRule*, *typicalExample*, and *Remove2*.

In the second experiment, we removed from the grammar the symbols responsible for sophisticated pruning techniques: *PostProcess* and *PrePruneRule*. As explained in Chapter 2, pruning is one of the optional elements in rule induction algorithms. Although it is present in virtually all the newer rule induction algorithms, we wanted to evaluate how the lack of pruning would impact the evolved rule induction algorithms. Note that simpler ways of pruning rules, such as the ones provided by the nonterminal *StoppingCriterion*, remained in this version of the grammar.

At last, in the third experiment, we modified the grammar to force the system to produce only bottom-up rule induction algorithms by removing the symbols *emptyRule* and *MakeFirstRule* from the grammar. The great majority of rule induction algorithms uses a top-down approach. So the main objective of this experiment was to find out how exclusively bottom-up algorithms would perform in the datasets in the meta-test set.

All the experiments reported in this section were run with the same 20 datasets used in the previous experiments discussed so far in this chapter. The GGP system's parameters are also the same ones defined in Section 6.2.3: population of 100 individuals, evolved for 30 generations, tournament size of 2, crossover rate of 0.7, mutation rate of 0.25, and reproduction rate of 0.05. Table 6.10 shows the predictive accuracies obtained by the GGP-RIs in the meta-test set when using the original version of the grammar (from now on referred as GGP-RIs-Original) followed by the results of the three grammar variations described above, named GGP-RIs-Basic, GGP-RIs-NoPrune and GGP-RIs-BottomUp, respectively. As in the other tables of this chapter, cells in dark gray represent results in which the GGP-RIs-Original obtained significantly better results than the other variations according to a paired two-tailed Student's t-test with 0.05 significance level. In contrast, cells in light gray represent a significantly better result of a grammar variation over the original grammar.

As can be observed in Table 6.10, the GGP-RIs-Basic obtain significantly better accuracies than the GGP-RIs-Original in two datasets, namely *segment* and *splice*. As this basic version of the grammar is simpler than the original one, we might conclude that it produced simpler rule induction algorithms, which performed better in these datasets. In the case of *segment*, one possible explanation for the results is the high degree of noise present in the data, as pointed out in [2]. In contrast, when looking at the accuracies obtained by the GGP-RIs-NoPrune, we notice that they obtained significantly worse results than the GGP-RIs-Original in the datasets *mushroom* and *splice*. Finally, the GGP-RIs-BottomUp obtained significantly worse results than the GGP-RIs-Original in half the datasets, and competitive results in the other half. Note that all the 25 rule induction algorithms produced by the GGP-Original followed the top-down approach. According to these results, in the datasets *crx*, *heart-c*, *ionosphere*, *monks-1*, and *mushroom* the top-down and bottom-up algorithms obtained the same level of predictive accuracy. The top-down approach, however, was more suitable for the other datasets.

In summary, it is not simple to compare different versions of the grammar used for guiding the search of the GGP system, as the system adapts to the grammar and produces different types of rule induction algorithms according to the functionalities available. This section presented results which demonstrated that a simpler version of the grammar can be as effective as a more sophisticated one, and explained how the grammar can be manipulated to produce a particular kind of rule induction algorithm.

6.2.6 GGP Versus Grammar-Based Hill-Climbing Search

Chapter 5 proposed a GGP system to search for accurate and innovative rule induction algorithms. The GGP system was first chosen as the search method for this task because of its global search nature, its intrinsic population-based parallelism and its basic characteristic of being a promising "automated invention machine." Besides, the use of a grammar as a way to provide available background knowledge about the target problem and guide the search seemed appropriate in the case of automatically evolving a rule induction algorithm.

As explained before, in GGP systems the grammar is responsible for determining the size of the search space. In the problem of automatically evolving rule induction algorithms, the grammar presented in Table 5.1 represents a search space of approximately five billion candidate rule induction algorithms (for details, see [17, p.196]). As shown in the results of the previous sections, the GGP system met our expectations in finding rule induction algorithms competitive with human-designed ones. But is this GGP system a good way to automatically search for rule induction algorithms? Or could a simpler search method obtain the same kinds of results?

In order to find answers to these questions we implemented a hill-climbing (HC) search method. The central idea of this hill-climbing search is the following [23]. It randomly generates one solution to a specific problem (which we call current solution), evaluates it, and then modifies it (a modification is equivalent to a mutation operation in the GGP system). If the new, modified solution is better than the current solution, the former replaces the latter. In contrast, if the new solution is equal to or worse than the current one, it is discarded and the current solution (which remains unchanged) is modified again. This process is carried out until a maximum number of solutions are evaluated.

In order to use this HC method to automatically evolve rule induction algorithms (which is comparable with the proposed GGP system), we randomly generate a rule induction algorithm by following the production rules of the grammar, as in the GGP system. The rule induction algorithm is evaluated using the meta-training set (see Section 6.2.1) and the same fitness used by the GGP system. It is then mutated and, if the new rule induction algorithm is better than the current one, the former replaces the latter. Otherwise the unchanged solution (the rule induction algorithm before the mutation) undergoes a new mutation operation. This process is repeated 3,000 times (which is equivalent to running the GGP system with 100 individuals for 30 generations, parameter values used in all GGP experiments reported in the previous sections). We call this method GHC (Grammar-based Hill Climber).

Therefore, both the GGP and the GHC systems use the same grammar, the same individual representation, and the same fitness function and evaluate the same number of candidate rule induction algorithms, making the comparison between the two systems as fair as possible. The two systems differ in that (a) the GGP system works with a population of candidate solutions, whereas the GHC system works with just one candidate solution at a time; (b) as the GGP system works with a population, individuals undergo a selection procedure before being modified, while in the GHC system an "elitist" strategy is used, and only the best candidate solution is preserved (in other words, the GHC system performs a local search, while the GGP system performs a global search); and (c) the GGP system creates new solutions via crossover and mutation whereas the GHC system uses only "mutation," which is implemented by exactly the same mutation operator used by GGP system.

Table 6.11 shows the results obtained by the described GHC method for the datasets in the meta-test sets. The tables show the name of the datasets followed by the predictive accuracies obtained by the GGP-RIs and the rule induction algorithms produced by the GHC method (denoted GHC-RIs) in five different runs (with different random seeds) of each method, respectively. As in the other tables presented in this chapter, all the results were obtained using a five-fold cross-validation procedure, and the numbers after the symbol "$\pm$" are standard deviations.

Results were compared using a paired two-tailed Student's t-test with significance level 0.01, and cells in dark gray represent significant wins of the GGP-RIs against the GHC-RIs. In Table 6.11, the GGP-RIs obtained significantly better results than the GHC-RIs in five out of ten cases. There was no case in which the GHC-RIs obtained significantly better results than the GGP-RIs. Hence, in general, we can say that the GGP system was much more effective in finding good rule induction algorithms than the GHC system.

Looking at the results reported in Table 6.11, the accuracies obtained by the GHC-RIs for the datasets *promoters* and *splice* are surprisingly low: while the GGP-

Table 6.11 Comparing the predictive accuracies (%) of the GGP-RIs and the GHC-RIs in the meta-test set

Dataset	GGP-RIs	GHC-RIs
crx	77.46±3.8	82.66±1.14
heart-c	76.72±1.5	78.75±1.03
ionosphere	87.04±2.2	84.64±1.97
monks-1	99.93±0.07	99.82±0.18
mushroom	99.99±0	99.03±0.07
promoters	78.98±2.93	60.26±1.96
segment	95.06±0.26	88.53±1.03
sonar	72.34±1.91	64.92±1.12
splice	88.68±0.31	65.2±0.27
wisconsin	95.58±0.74	93.56±0.56

RIs obtained accuracies of 78.98 ± 2.93 for *promoters* and 88.68 ± 0.31 for *splice*, the GHC-RIs obtained accuracies of 60.26 ± 1.96 and 65.2 ± 0.27 for *promoters* and *splice*, respectively. Analyzing the actual rule induction algorithms evolved, we notice that two out of five algorithms produced by the GHC system (in the five different runs varying the random seed) were responsible for the very low accuracy. These two algorithms produced unordered rule sets (rather than ordered rule lists), and post-processed them by removing one or two conditions at a time from the final rule set. These strategies were not very successful.

Figure 6.1 shows a comparison of the evolution of candidate rule induction algorithms along the GGP and GHC systems' searches. The plot shows fitness values against the number of evaluations produced in one run of each method, where in both runs the same random seed was used to initialize the candidate solutions of the corresponding method. Note that while for the GHC system we have results for all the evaluations (since the GHC system works with one candidate solution at a time), for the GGP system we show the fitness of the best individual at the end of each generation (a multiple of 100 evaluations), since the GGP system is a population-based method.

Observe that in both searches the values of the fitness do not monotonically increase as the number of evaluations increase (as would be expected in a "static" fitness function scenario). This is because, as explained before, in order to avoid overfitting, at each generation of the GGP system (which is equivalent to 100 evaluations in the GHC system's search), for each dataset in the meta-training set, the training and validation subsets were changed, which effectively means the fitness landscape changes at every 100 individual evaluations. It is very interesting to notice the effect this design issue has on both algorithms.

Also, observe that there is a number of points in which the GHC system has a better solution than the GGP system. One of these points is indicated in Fig. 6.1 by a dashed vertical line. Notice that the dashed vertical line is in the evaluation number 2,700, i.e., one evaluation before the datasets in the meta-training sets were randomly redivided into training and validation subsets. In other words, as the GHC system improves a solution over the same set of data 100 times, it probably overfits

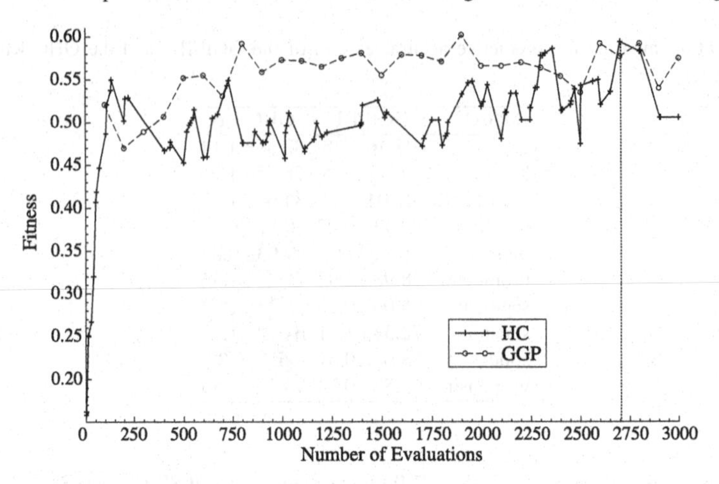

Fig. 6.1 Evolution of the GGP system versus the GHC system

the rule induction algorithm to the data, which explains the better and better results from evaluations 2,601 to 2,700 in Fig. 6.1, followed by a progressive drop from evaluation 2,701.

It is also worth noting that out of the 3,000 mutation operations performed by each run of the GHC system, on average only 93 were successful in producing better rule induction algorithms than the previous one (i.e., the "parent" algorithm) in five runs. This number tell us that only $\simeq 3\%$ of the mutation operations improved the current candidate solution. Unfortunately, we cannot do this same kind of analysis with the GGP system. The best way to compare the improvement of the GGP system's individuals would be to check if the individuals produced after crossover and mutation operations were better than their parents. However, as the data changes at each generation, parents and their offspring are evaluated in different sets of data, and are not directly comparable.

In conclusion, while the GGP system has the feature of searching in parallel (evaluating 100 individuals in each generation and breeding them), with the associated advantages of performing a global search, the GHC system has the feature of performing a sequential stepwise search, which is simpler but constitutes a form of local search, intuitively with considerably less exploratory power than the GGP system. When it comes to search for rule induction algorithms, the GGP system produced better results than the GHC system.

6.2.7 MOGGP: A Multiobjective Version of the Proposed GGP

One of the motivations to automatically evolve rule induction algorithms is the simplicity and interpretability of the classification models they generate. This is in con-

Table 6.12 Comparing the MOGGP-RIs to other rule induction algorithms, taking into account both predictive accuracy and number of conditions in the produced rule sets or lists according to the concept of Pareto dominance

	Neutral	Dominates	Dominated
MOGGP-RIs *vs* GGP-RIs	4	6	0
MOGGP-RIs *vs* Baseline Algorithms	7	23	0

trast with more mathematically sophisticated algorithms such as support vector machines, which usually produce a "black-box" model, hardly interpretable by the user. However, in the first experiments performed in the previous sections, the simplicity of the models generated by the evolved rule induction algorithms was not considered when evaluating those rule induction algorithms.

As explained in Section 5.5.3, we developed a more sophisticated version of the GGP system that takes into account in its fitness function both the predictive accuracy of a rule induction algorithm and the total number of conditions belonging to all rules in the classification model built by that algorithm. These two fitness criteria are simultaneously taken into account using the concepts of multiobjective optimization and Pareto dominance, and this new version of the GGP system is denoted MOGGP [20].

This section presents the GGP-RIs evolved by the MOGGP, denoted MOGGP-RIs. In contrast with the experiments run with the single-objective version of the GGP, here the crossover and mutation rates were not optimized, and the MOGGP was run with the same sets of parameter values found in the experiments reported in Section 6.2.1. The justifications for using the same parameter values optimized for the single-objective GGP system are twofold. First, as concluded in Section 6.2.1, the single-objective GGP system was not sensitive to crossover and mutation rate variations. Secondly, in order to compare the results of the single-objective and multiobjective versions of the GGP system and draw conclusions concerning the advantages of a multiobjective approach, it is fair to compare their results when they are run with exactly the same parameter values.

Therefore, we kept all the parameter values for the MOGGP exactly the same as those used for the single-objective GGP system: population size of 100, evolved for 30 generations, using a tournament size of two, and crossover, mutation, and reproduction rates of 0.7, 0.25, and 0.05, respectively.

In order to compare the MOGGP-RIs with other rule induction algorithms, taking into account both the accuracy and the rule size they obtained, we used an adapted version of the Pareto dominance concept, which also considers statistically significant differences. This adapted Pareto dominance concept states that a solution S_1 dominates a solution S_2 if two conditions are satisfied. First, if every objective value of S_1 is not statistically significantly worse than the corresponding objective value in S_2. Secondly, if at least one of the objective values of S_1 is statistically significantly better than the corresponding objective value of S_2 (statistical significance is determined by the results of a paired two-tailed Student's t-test with significance level 0.05).

Table 6.13 Predictive accuracies (%) and number of conditions in the rule sets or lists obtained by the MOGGP-RIs and the GGP-RIs for datasets in the meta-test set

Dataset	MOGGP-RIs		GGP-RIs	
	Accuracy	Size	Accuracy	Size
crx	83.33±1.26	13.52±0.72	77.46±3.8	99.4±5.98
segment	92±0.67	25.64±1.22	95.06±0.26	83.2±6.13
sonar	68.04±1.74	4.6±0.75	72.34±1.91	20.2±1.32
heart-c	76.46±1.82	7.2±0.9	76.72±1.5	50.6±2.92
ionosphere	85.48±1.63	7.88±0.62	87.04±2.2	24.2±1.53
monks-1	99.78±0.22	11.64±0.39	99.93±0.07	13±2.05
mushroom	99.66±0.22	15.16±0.38	99.98±0.02	15.2±0.58
wisconsin	92.1±0.71	9.68±0.57	95.58±0.74	48±10.62
promoters	71.84±5.24	3.96±0.38	78.98±2.93	14.6±2.27
splice	87.68±0.5	42.52±2.3	88.68±0.31	271.8±12.02

Table 6.14 Predictive accuracies (%) and the number of conditions in the rule sets or lists obtained by the baseline algorithms for datasets in the meta-test set

Dataset	OrdCN2		UnordCN2		C45Rules	
	Accuracy	Size	Accuracy	Size	Accuracy	Size
crx	80.16 ± 1.27	101.4 ± 3.46	80.6 ± 0.93	101.6 ± 2.38	84.82 ± 1.53	34 ± 1.38
segment	95.38 ± 0.28	73.8 ± 1.74	85.26 ± 0.87	102.8 ± 2.15	88.16 ± 7.72	96.8 ± 12.71
sonar	70.42 ± 2.66	19.4 ± 0.87	72.42 ± 1.4	50.2 ± 3.02	72.4 ± 2.68	14.6 ± 4.3
heart-c	77.9 ± 1.96	37.2 ± 1.24	77.54 ± 2.85	70 ± 3.54	74.2 ± 5.43	22 ± 5.63
ionosphere	87.6 ± 2.76	20.2 ± 1.53	90.52 ± 2.03	37 ± 1.84	89.06 ± 2.71	10.2 ± 4.34
monks-1	100 ± 0	11 ± 0.71	100 ± 0	61 ± 0	100 ± 0	61 ± 0
mushroom	100 ± 0	15.6 ± 0.24	100 ± 0	26 ± 0	98.8 ± 0.06	18.6 ± 2.73
wisconsin	94.58 ± 0.68	32.6 ± 1.36	94.16 ± 0.93	53.8 ± 2.91	95.9 ± 0.56	19.2 ± 0.86
promoters	81.9 ± 4.65	10.4 ± 0.75	74.72 ± 4.86	23.6 ± 1.36	83.74 ± 3.46	10.8 ± 1.02
splice	90.32 ± 0.74	256.2 ± 5.08	74.82 ± 2.94	172.6 ± 9.75	89.66 ± 0.78	119.8 ± 29.68

Table 6.12 presents the number of classification models produced by the MOGGP-RIs that neither dominate nor are dominated by the classification models produced by the GGP-RIs or the baseline algorithms (*Neutral* column), the number of models produced by the GGP-RIs or the baseline algorithms that the MOGGP-RIs dominate (*Dominates* column), and finally the number of models produced by the MOGGP-RIs that are dominated by a model produced by the GGP-RIs or a baseline algorithm (*Dominated* column). This table was built by applying the statistical significance-adapted Pareto dominance concept to the results presented in Tables 6.13 and 6.14.

It is important to emphasize that the results in Tables 6.13 and 6.14 are reported here only for the sake of completeness, as our objective is to generate rule induction algorithms where both accuracy and rule model comprehensibility are equally important. This approach is suitable particularly for applications where, in practice, the classification model will not be directly used to predict the class of individual examples, but will rather be interpreted by a user who is an expert in the application domain and the data, in order to try to gain new insights about the domain or the data. There are several types of applications where classification models are often induced mainly to be interpreted by users, such as biomedical applications and bioinformatics [15, 3, 21]. In such applications, discovering comprehensible knowledge is desirable for several reasons [24, 9], such as for increasing the confidence of the user in the system's results, leading to new insights about the data and the formulation of new hypotheses, and for detecting errors in the data. For instance, several new insights about the application domain provided by a comprehensible classification model are discussed in [13] in the context of protein function prediction. As another example of the usefulness of comprehensible models, even if the predictive accuracy is not very good, Wong and Leung [26] discovered classification rules with a relatively low or moderate degree of accuracy (around 40–60%) that were considered, by senior medical doctors, novel and more accurate than the knowledge of some junior doctors.

Table 6.13 presents the predictive accuracies and the number of conditions per rule set or rule list obtained by the MOGGP-RIs and GPP-RIs, while Table 6.14 presents the results found by the baseline algorithms. These results were obtained using a five-fold cross-validation procedure, and compared using a paired two-tailed Student's t-test with significance level 0.05, performed over five independent runs of the GGP system for each data fold. Cells in dark gray represent statistically significant wins of the MOGGP-RIs against the other algorithms, while light gray cells represent the MOGGP-RIs' statistically significant losses. Note that, although the MOGGP system finds a set of non-dominated solutions, in practice we want to select a single solution (rule induction algorithm) to compare it with the baseline algorithms. Instead of manually choosing a single rule induction algorithm among the non-dominated ones, to keep the experiments as automated and unbiased as possible we have also automated that selection. That is, the MOGGP returns the non-dominated rule induction algorithm (in the last generation) having the largest value of the f_{tb} function (see Section 5.5.3).

To illustrate the logic behind the results presented at Table 6.12, let us consider the cases of *crx* and *monks-1* with ordered CN2. In both cases the accuracies of the MOGGP-RIs and ordered CN2 are competitive – i.e, the differences between the predictive accuracies of MOGGP-RIs and ordered CN2 classification models in those two datasets are not statistically significant. However, in *crx* the classification models generated by the MOGGP-RIs have significantly smaller numbers of rule conditions than the model generated by ordered CN2, and so we say that the MOGGP-RIs dominate ordered CN2. A different situation occurs for *monks-1*, where the size of the models generated by both the MOGGP-RIs and ordered CN2 are also competitive, and so we say that the MOGGP-RIs and ordered CN2 have a neutral relationship. Finally, the MOGGP-RIs are said to be dominated by another algorithm if they are significantly worse in one objective (accuracy or model size) and not significantly better in another. In other words, if the MOGGP-RIs obtain a significantly smaller accuracy (or larger rule set) than the algorithm in question, and at the same time do not discover a significantly smaller rule set (larger accuracy), then they are said to be dominated by the respective algorithm.

An analysis of statistical significance-based Pareto dominance taking into account both the accuracy and the size of the models produced shows that in four out of ten cases the MOGGP-RIs and the GGP-RIs present a neutral relationship, while in the other six cases the MOGGP-RIs' models dominate the GGP-RIs' models. In contrast, when comparing the MOGGP-RIs with the baseline (human-designed) algorithms, in 23 out of 30 cases (three baseline algorithms times ten datasets) the MOGGP-RIs' models dominate the baseline algorithms' models, and the former are never dominated by the latter. For a more detailed analysis of these results and the algorithms produced by the MOGGP approach, the reader is referred to [20].

As a final remark relevant to the topic of this section, it should be recalled that, as an alternative to the multiobjective approach based on Pareto dominance discussed above, it is certainly possible to take a different perspective and use the lexicographic approach, as discussed in Section 3.3.2. The crucial difference between the two approaches is that they are based on different perspectives about the relative importance of different objectives to the user. As discussed above, the Pareto approach seems appropriate when the user does not have any clear idea about the relative importance of quality criteria such as the predictive accuracy and the size of a classification model (rule set, in our case), i.e., when the user does not even know which of those criteria is more important than the other. However, if the user considers that, say, maximizing predictive accuracy is more important than minimizing rule set size (a position that is taken by most data mining researchers), then the lexicographic approach would be a more natural choice, since it can directly take this kind of user preference into account, unlike the Pareto approach, as discussed in Section 3.3.2. Hence, an interesting research direction would be to extend the proposed GGP system to perform multiobjective optimization based on the lexicographic approach.

6.2.8 A Note on the GGP System's Execution Time

Formal analyses of the computational time complexity of evolutionary algorithms are not very often performed. Although theoretical studies [12] suggest the use of Markov chains models in order to perform these analyses, this is not a simple task.

Performing this kind of theoretical analysis is out of the scope of this book. Instead, we discuss the factors which tend to define the best and worst case scenarios in a run of the proposed GGP system.

The most time consuming operation in the GGP system is the evaluation of the individuals. As each individual represents a complete rule induction algorithm, the GGP system's overall runtime depends on the rule induction algorithms' overall runtime.

More precisely, the runtime of the fitness function of the GGP system (by far the most time consuming part of the system) depends on three main factors:

- The number of datasets in the meta-training set.
- The number of attributes (and attribute values, in the case of nominal attributes) and examples in each dataset in the meta-training set.
- The rule induction components (incorporated in the grammar) used by the GGP system to produce a complete rule induction algorithm.

Table 6.15 GGP system's runtime for different experiments' configurations when training the GGP system with a different number of datasets in the meta-training set

Meta-training set	Time (hr:min)	
# Datasets	Best	Worst
6	33:21	48:29
8	25:39	32:48
10	18:29	34:02
12	33:12	53:30
14	40:31	64:21

The third factor, in particular, considerably changes the rule induction algorithms' runtimes. This is because, when analyzing the symbols of the grammar, we noticed that a few of them perform operations with a time complexity higher than linear. *Add2*, for instance, considers all the combinations of attribute-value pairs two by two, so its time complexity is quadratic with respect to the number of attributes. As another example, *RemoveCondRule* post-prunes rule models by removing one or two conditions at a time from each rule in the rule model.

Since it is so difficult to estimate a best and worst runtime for the GGP system, we report here the best and worst runtimes empirically obtained for each of the experiments with the GGP system reported in this chapter. All the experiments were performed on Pentium 4 duo processor machines with 1 GB RAM and running Linux.

Table 6.15 shows the results of the best and worst case scenarios when running the proposed GGP system with different numbers of datasets in the meta-training set. The first column presents the number of datasets used in the meta-training set, followed by the best and worst runtimes, reported in the format hours:minutes. Note that each runtime in Table 6.15 refers to the time taken to compute a single run of the GGP system.

It is important to remark that the execution time of the rule induction algorithms can still be improved, and their code is not optimized. As the GGP system considers many combinations of nonterminals and terminals, it is not simple to write the code for all the grammar symbols in a way that all the combinations run in the best possible time.

6.3 Evolving Rule Induction Algorithms Tailored to the Target Application Domain

As explained before, the GGP system proposed in Chapter 5 can be used in at least two different frameworks: to evolve rule induction algorithms that are robust across different application domains or to evolve rule induction algorithms tailored to a specific application domain. This latter approach is the one studied in this section.

There are two main differences in the setup of the experiments in this section when compared to the experiments in Section 6.2. First, while in Section 6.2 a set of datasets was used in the GGP system's meta-training set, in this section only one dataset is used in the GGP system's meta-training set. Secondly, while in Section 6.2 the meta-test set was composed of datasets coming from different application domains to the datasets present in the meta-training set, in this section the meta-test set contains a dataset coming from the same application domain as the dataset in the meta-training set.

For all the experiments run in this section, each dataset was divided into three subsets: a training, a validation, and a test set. The training and validation sets were inserted into the GGP system's meta-training set and used to evaluate the candidate rule induction algorithms found by the GGP system. The meta-test set was not used during the evolution of the GGP system, being reserved only for evaluating the predictive accuracy of the evolved rule induction algorithm. In order to evaluate this predictive accuracy, after the GGP system was run, the training and validation sets used in the meta-training set during evolution were merged to create a new training set. The evolved rule induction algorithm used this new training set to create a rule set (classification model), which was then evaluated on the unseen test set.

For this set of experiments, optimizing all the GGP system's parameters for each specific dataset was not a practical option, since the parameter set would have to be optimized for each dataset. Hence, we used the same parameters used in the experiments presented in Section 6.2: population size of 100, evolved in 30 generations, and tournament size of two. Regarding the crossover and mutation rates, we used the value 0.7 for crossover and 0.25 for mutation, the reproduction rate being 0.05. These values were chosen because they were considered the most suitable ones in a previous set of experiments (see Section 6.2.1).

For each dataset, the GGP system was run 25 times. As before, we had five different random seeds, but in order to obtain more statistical support for the results, for each random seed, we evolved five GGP-RIs. In each of these five runs, we varied the data in the meta-training and meta-test sets. In order to do that, the entire dataset was divided into five partitions and, as in a conventional five-fold cross validation procedure, each time three data partitions were used for training, one for validation and one for test.

Regarding the datasets used in the experiments, they were divided in two groups. First, we ran experiments for the 20 datasets used in the experiments in Section 6.2, as reported in Section 6.3.1. Second, we ran experiments for five bioinformatics datasets, as reported in Section 6.3.3.

6.3.1 Experiments with Public UCI Datasets

Initially, we ran experiments using the GGP system to evolve rule induction algorithms tailored to a specific dataset taken from the UCI (University of California at Irvine) dataset repository [16]. Table 6.16 shows the predictive accuracies obtained by the GGP-RIs (rule induction algorithms evolved by the GGP system) followed by the values of the predictive accuracies of the baseline algorithms, namely ordered CN2 and unordered CN2, Ripper, and C4.5Rules. Results were compared using a paired two-tailed Student's t-test with significance level 0.05. Cells in dark gray represent statistically significant wins of the GGP-RIs over the respective baseline method, while cells in light gray represent statistically significant wins of the respective baseline method against the GGP-RIs. Note that, for the dataset *splice*, an attribute selection method was applied before the dataset was given to the GGP system. The reason for that was the computational time required to design rule induction algorithms for this dataset. *splice* has 63 attributes and 3,190 examples, and training the GGP system with it was extremely slow.

The attribute selection method applied to *splice* is not optimal, as attribute selection is out of the scope of this book. It followed the filter approach, and was based on the attributes' information gain ratios [25]. In this approach, the attributes were ranked according to their gain ratio, and then removed, ten by ten, from the worst to the best (according to their rank), from the original dataset. Every time ten attributes were removed from the training set, the new dataset was mined by the four baseline algorithms in a five-fold cross-validation procedure (i.e, both versions of CN2, Ripper, and C4.5Rules). This process of removing ten attributes from the training set was repeated until the predictive accuracy of at least one of baseline algorithms dropped. In this case, the last ten attributes removed were added again to the training set and then removed, one by one, from the worst to the best, while the predictive accuracy of the classifiers did not decrease. As a result of the attribute selection method, 14 out of 63 attributes were used. We emphasize that attribute selection was performed only for the *splice* dataset.

Results in Table 6.16 show that the GGP-RIs obtain predictive accuracies significantly better than the baseline algorithms in ten out of 80 cases, and accuracies significantly worse than the baseline algorithms in five cases. In the remaining 65 cases, the GGP-RIs' accuracies are considered statistically competitive with the ones generated by the baseline human-designed rule induction algorithms.

Table 6.16 presents six datasets that have not appeared before in the set of meta-test sets, namely *balance-scale*, *monks-2*, *vehicle*, *vowel*, *glass*, and *lymphs*. Seven out of the ten cases where the GGP-RIs obtain significantly better accuracies than a baseline method occur in these datasets. The other three cases reflect the results of previous experiments with a set of datasets in the meta-training set, and appear in *mushroom*, *segment*, and *splice*. In these cases the accuracies of C4.5Rules and unordered CN2 are too low when compared to the other methods. The five significant losses of the GGP-RIs occur in the datasets *hepatitis*, *lymph*, *vehicle*, *vowel*, and *splice*.

These results show that apparently the GGP system was not able to find ordered CN2, the algorithm with best performance among the four manually designed algorithms in Table 6.16 for the dataset *lymph*, although ordered CN2 is included in the GGP system's search space. However, when analyzing the 25 GGP-RIs produced for *lymph*, we observed that six of them were actually instances of ordered CN2.

Table 6.16 Predictive accuracy rates (%) for GGP-RIs tailored to a specific dataset

Dataset	GGP-RIs	OrdCN2	UnordCN2	Ripper	C45Rules
balance-scale	79.77±1.11	81.62 ± 1.53	77.1 ± 1.18	74.04 ± 1.27	76 ± 1.58
crx	82.22±1.51	80.16 ± 1.27	80.6 ± 0.93	84.37 ± 1.21	84.82 ± 1.53
glass	65.36±1.89	68.44 ± 4.58	69.42 ± 2.27	66.13 ± 3.32	67.72 ± 4.23
heart-c	78.67±1.37	77.9 ± 1.96	77.54 ± 2.85	77.53 ± 1.1	74.2 ± 5.43
hepatitis	80.02±1.07	81.94 ± 5.02	83.34 ± 1.83	86.03 ± 1.14	83.36 ± 0.9
ionosphere	85.65±2.85	87.6 ± 2.76	90.52 ± 2.03	89.61 ± 1.75	89.06 ± 2.71
lymph	75.85±2.04	82.44 ± 1.7	80.36 ± 4.06	79.03 ± 4.62	81.42 ± 4.05
monks-1	100±0	100 ± 0	100 ± 0	93.84 ± 2.93	100 ± 0
monks-2	89.67±1.22	87.26 ± 1.09	76.5 ± 0.83	64.1 ± 0.8	73.78 ± 2.25
monks-3	98.38±0.6	97.46 ± 0.74	99.1 ± 0.4	98.54 ± 0.46	94 ± 4.89
mushroom	100±0	100 ± 0	100 ± 0	99.96 ± 0.04	98.8 ± 0.06
pima	73.57±0.53	69.34 ± 2.13	74.6 ± 0.38	73.91 ± 1.65	71.04 ± 1.67
promoters	76.07±3.36	81.9 ± 4.65	74.72 ± 4.86	78.18 ± 3.62	83.74 ± 3.46
segment	95.55±0.25	95.38 ± 0.28	85.26 ± 0.87	95.44 ± 0.32	88.16 ± 7.72
splice	89.53±0.65	90.32 ± 0.74	74.82 ± 2.94	93.88 ± 0.41	89.66 ± 0.78
sonar	70.9±2.27	70.42 ± 2.66	72.42 ± 1.4	72.88 ± 4.83	72.4 ± 2.68
vehicle	68.34±0.95	70.16 ± 1.68	59.68 ± 2.01	66.46 ± 1.94	71.94 ± 1.29
vowel	73.66±1.11	76.64 ± 0.93	62.02 ± 1.63	68.93 ± 2.64	58.68 ± 6.24
wisconsin	94.27±1.1	94.58 ± 0.68	94.16 ± 0.93	93.99 ± 0.63	95.9 ± 0.56
zoo	91.85±1.87	92.64 ± 1.33	92.52 ± 2.21	89.47 ± 1.66	92.56 ± 1.45

As explained before, apart from measuring the predictive accuracy obtained by the GGP-RIs, it is also important to know whether the GGP system is capable of producing innovative rule induction algorithms, which are not simply a reproduction of the human-designed ones. The experiments showed in Table 6.16 generated nothing less than 500 evolved algorithms (20 datasets times 25 algorithms per dataset). The first point we noticed when analyzing the algorithms evolved with single datasets in the meta-training set is that they are much more original (different from manually designed algorithms) than the ones produced when using a set of datasets in the meta-training set. We can say that approximately 25% of the algorithms produced were instances of ordered CN2, or just changed one or two of its main components. Nevertheless, there were plenty of algorithms using a more creative way of finding rules.

From that, we can conclude that when evolving rule induction algorithms for a set of datasets, the GGP system is more cautious and gives preference to simpler algorithms, which are more likely to perform well across a set of very different datasets. However, when tailoring the evolved rule induction algorithm to the data, more "specialized" algorithms were generated.

Another point to be noticed is that, for the same random seed used to create the initial population of the GGP system, the five evolved algorithms generated with different sets of data during a cross-validation process usually follow the same broad strategy. For instance, in the dataset *balance-scale*, four out of the five algorithms use a bottom-up approach instead of a top-down one. In addition, three out of these four create the first rule using the typicality measure, while the fourth one starts the search with a random example.

Algorithm 6.4: Example of a decision list algorithm created by the GGP system specifically for the dataset *crx*

```
Divide the training data in Build and PostPrune
RuleList = ∅
repeat
    Divide the Build data in Grow and PrePrune
    bestRule = rule created from a typical example
    candidateRules = ∅
    candidateRules = candidateRules ∪ bestRule
    while negative examples covered by bestRule ≠ ∅ do
        for each candidateRule CR do
            newCandidateRules = ∅
            Remove 2 conditions at a time from CR
            Evaluate CR using its information content in Grow
            if CR better than bestRule then
                bestRule = CR
            newCandidateRules = newCandidateRules ∪ CR
        candidateRules = best 10 rules selected from newCandidateRules
    bestRule' = Remove the last condition from bestRule
    Evaluate bestRule' into PrePrune using Laplace estimation
    if bestRule' better than bestRule then
        bestRule = bestRule'
    RuleList = RuleList ∪ bestRule
    Remove from Build data all examples covered by bestRule
until at least 99% of the examples in the training set are covered
RuleList' = RuleList
for each rule R in the RuleList' do
    notImproving = false
    repeat
        Remove 2 conditions at a time from R
        Evaluate RuleList' in PostPrune
        if laplaceEstimation(RuleList') ≥ laplaceEstimation(RuleList) then
            RuleList = RuleList'
        else notImproving = true
    until notImproving
```

As mentioned earlier, the experiments whose results are reported in Table 6.16 generated 500 GGP-RIs. Here we are going to show two GGP-RIs, selected out of those 500 GGP-RIs. The selection criteria were based on both the "originality" of the algorithm (which is, to a large extent, a subjective criterion) and its predictive accuracy in the test set, although the second criterion does not have much statistical support, as it is based on only one run of the algorithm (corresponding to a single test set, i.e., a single fold of the cross-validation procedure). Algorithm 6.4 was evolved for the dataset *crx*. It produces the initial rule based on a typical example extracted from the training set (details on how to select the typical example can be found in Section 5.2), and removes two conditions at a time from it. The generated candidate rules are evaluated using their information content, and rules are refined until the best rule found so far does not cover any negative examples. Rules can be pre-pruned before being inserted into the rule list by removing the last condition inserted into them. The rule production process is carried out until 99% of the examples in the training set are covered. A post-pruning phase also allows the algorithm to remove two conditions at a time from each rule in the final rule list, as long as the value of the Laplace estimation in the post-pruned rule set is higher than the value of the Laplace estimation in the original rule list. The algorithm works with three sets of data, one to grow, one to pre-prune, and a third to post-prune the rules.

This algorithm is innovative in the sense that it starts the search with a typical example (instead of a random one) and removes two conditions at a time from it, accounting for some attribute interaction. Recall that the selection of a typical example, based on the principles used in instance-based learning algorithms, is an innovative feature of the grammar that is not found in any manually designed rule induction algorithm, to the best of our knowledge. The fact that the rules are refined until no negative examples are covered could lead to overfitting. However, the generalization capability of the classifier is secured by both pre- and post-prune phases. First the rules are pre-pruned by removing the last condition added to them, and evaluated using the Laplace estimation this time. Later, a post-processing phase helps to ensure a compact rule list.

Algorithm 6.5 shows a GGP-RI produced for the dataset *ionosphere*. The algorithm produces a rule set using a top-down strategy. The novelty of this algorithm lies in the way rules are refined, depending on how many conditions they have. Two conditions at a time are added to rules with less than five conditions, while only one condition at a time is added to rules with at least five conditions. A beam search of size 3 is performed, and rules are evaluated using accuracy. It excludes rules with less than 80% accuracy from the set of candidates. Rules are produced for a given class until 97% of the examples of the current class are covered.

6.3.2 GGP-RIs Versus GHC-RIs

As we did for the experiments in Section 6.2.6, we compared the predictive accuracies obtained by the GGP-RIs (rule induction algorithms designed by the GGP system) with the ones obtained by the rule induction algorithms designed by the grammar-based hill-climbing system, denoted GHC-RIs. Table 6.17 shows the results of these comparisons. Cells in dark gray represent cases where the GGP-RIs obtained predictive accuracies significantly better than the GHC-RIs according to a paired two-tailed Student's t-test with significance level 0.01.

As illustrated in Table 6.17, the GGP-RIs obtained significantly better accuracies than the GHC-RIs in ten out of 20 cases. In three out of these ten cases, namely *vowel*, *vehicle*, and *glass*, the accuracies found by the GHC-RIs were surprisingly low. In the case of *glass*, the type of algorithms generated explains the poor accuracy. It was caused by the GHC-RIs, which used a pre-pruning method requiring a different set of data, and also a post-pruning method (also requiring a different set of data). As *glass* was trained with only 129 examples, the datasets reserved for pre- and post-pruning were not big enough to give statistical support to the results. We

Algorithm 6.5: Example of a rule set algorithm created by the GGP system specifically for the dataset *ionosphere*

```
RuleSet = ∅
for each class C in the training set do
    repeat
        bestRule = an empty rule
        candidateRules = ∅
        candidateRules = candidateRules ∪ bestRule
        while candidateRules ≠ ∅ do
            for each candidateRule CR do
                newCandidateRules= ∅
                if number of conditions in CR < 5 then
                    Add 2 conditions at a time to CR
                else
                    Add 1 condition at a time to CR
                Evaluate CR using accuracy
                if accuracy(CR) > 80% then
                    if CR better than bestRule then
                        bestRule = CR
                    newCandidateRules = newCandidateRules ∪ CR
            candidateRules = 3 best rules selected from newCandidateRules
        RuleSet = RuleSet ∪ bestRule
        Remove from training set all class C examples covered by bestRule
    until at least 97% of the examples of class C in the training set are covered
Class clashes when classifying new examples are solved using the LS content criterion
```

recognize that this division of data might cause problems in small datasets. However, the GGP system was able to overcome this problem, and only four out of the 25 GGP-RIs generated use a pre-pruning method that requires a different set of data, while none of those 25 algorithms use a post-pruning method. In contrast, the GHC system was not able to detect the problem.

An analysis of the GHC-RIs generated also showed that, in general, the algorithms designed by the GHC system when using one dataset in the meta-training set (in order to evolve a rule induction algorithm tailored to a single application domain) were also more original than the ones found by the GHC system when using a variety of datasets in the meta-training set (in order to evolve a more robust rule induction algorithm).

6.3.3 Experiments with Bioinformatics Datasets

In Section 6.3.1 we showed the results obtained by the GGP system when evolving rule induction algorithms tailored to specific datasets from the well-known UCI dataset repository [16]. In this section, we apply the GGP system to evolve rule induction algorithms tailored to each of the five bioinformatics datasets described in Table 6.18.

Table 6.17 Comparing the predictive accuracies (%) of the GGP-RIs and the GHC-RIs tailored to a specific dataset

Dataset	GGP-RIs	GHC-RIs
balance-scale	79.77±1.11	75.65±1.18
crx	82.22±1.51	80.5±0.87
glass	65.36±1.89	54.42±1.27
heart-c	78.67±1.37	78.72±0.88
hepatitis	80.02±1.07	78.66±0.8
ionosphere	85.65±2.85	83.43±1.32
lymph	75.85±2.04	73.47±1.71
monks-1	100±0	95.62±1.75
monks-2	89.67±1.22	69.53±1.2
monks-3	98.38±0.6	84.12±2.07
mushroom	100±0	97.31±1.43
pima	73.57±0.53	71.74±0.93
promoters	76.07±3.36	67.48±2.31
segment	95.55±0.25	87.96±0.71
splice	89.53±0.65	77.63±1.32
sonar	70.9±2.27	63.19±2.61
vehicle	68.34±0.95	56.65±4.25
vowel	73.66±1.11	42.88±3.35
wisconsin	94.27±1.1	92.32±0.78
zoo	91.85±1.87	89.6±0.83

In all these datasets, each example represents a protein. Proteins are produced from genes by the cells of biological organisms. Proteins are the main elements of the cell, and perform almost all the functions related to cell activity. Their basic structure (called primary structure in biology) consists of a linear sequence of amino acids, where each amino acid is composed of a short segment of DNA. Different types of proteins are associated with different biological functions, and there are a very large number of functions that can be performed by proteins, such as transporting important molecules (e.g., oxygen) throughout the body and catalyzing (speeding up) chemical reactions. Many diseases are caused by or at least related to some malfunctioning of proteins, so understanding protein functions is important for understanding, diagnosing, and treating diseases.

Ideally, we would like to know the function of every protein, given its primary sequence, i.e., its linear sequence of amino acids. However, although reading the linear sequence of amino acids of a protein is relatively easy with current technology, knowing the function performed by the protein is much more difficult. This is because, although a protein is produced in its basic form as a linear sequence of amino acids, a protein actually folds and takes a specific three-dimensional shape directly related to its function, and this folding process is very hard to predict. Protein folding prediction is a major open problem in molecular biology.

Instead of trying to predict the folding of a protein, which would involve predicting the position of each atom in the protein (a very large molecule), and then using the predicted three-dimensional shape of the protein to determine its function; it is possible to try to predict the function of a protein directly from its primary sequence

Table 6.18 Bioinformatics datasets used by the GGP system

Dataset	Examples	Attributes		Classes	Def. Acc.
		Nomin.	Numer.		(%)
Postsynaptic	4303	10(444)	0(2)	2	93.96
GPCR-Prosite	6261	30(128)	2(2)	9	75.16
GPCR-Prosite-L2	6162	25(128)	2(2)	50	33.7
GPCR-Prints	5422	24(282)	2(2)	8	77.8
GPCR-Interpro	7461	30(449)	2(2)	12	64.4

of amino acids. This is the approach that we follow in this section. In this approach, as said earlier, each example corresponds to a protein. The classes to be predicted are types of biological functions associated with the broad types of proteins in the dataset. One of the main challenges is how to define a good protein representation, i.e., how to define the predictor attributes that will describe the characteristics of each protein in the dataset.

In our experiments, each protein (example) is described mainly by a set of "motifs." A motif is a pattern or a "signature" typically found in some proteins. It is basically a sequence or partial sequence of amino acids that can be used to identify the function and/or the family of a protein [10]. Each motif is represented by a binary attribute, which indicates the presence or absence of the motif in each protein. This type of protein representation implicitly incorporates a lot of background knowledge about proteins available in the literature, since protein motifs have been created and refined by expert biologists and bioinformaticians for a long time.

Table 6.18 shows the number of examples, the number of attributes, and the number of classes for each bioinformatics dataset, followed by the default accuracy, i.e., the percentage of examples (proteins) belonging to the most frequent class in the dataset. Observe that, in the column "Attributes," the numbers in parentheses show the original number of attributes the datasets had before an attribute selection method was applied during a preprocessing step.

The datasets shown in Table 6.18 were preprocessed by using an attribute selection method for two reasons. First, due to the large number of predictor attributes in the original dataset, intuitively there are many attributes that are (to a large extent) irrelevant or redundant. The objective of attribute selection is to reduce the dimensionality of a dataset by identifying and removing irrelevant or redundant attributes without sacrificing predictive accuracy [14]. Secondly, the reduction in the dataset size makes the application of the GGP system much more efficient.

For the postsynaptic dataset (to be briefly described below) we used the top ten attributes selected by a particle swarm optimization algorithm [7], since these selected attributes are already available in the literature. Note that, in [7], a wrapper approach is used to select a set of optimal attributes for a naive Bayes classifier. We recognize that these attributes are not likely to be optimized for the algorithms considered in this work. Nevertheless, the issue of optimal attribute selection is out of the scope of this work and, as the GGP-RI is produced during the GGP system run, it is not practical to perform a customized attribute selection for each candidate rule induction algorithm beforehand.

For the other four datasets (of GPCRs, to be briefly described below), an attribute selection method based on the attributes' information gain ratios [25], as applied to *splice* in Section 6.3.1, was used. Recall that, in this approach, the attributes were ranked according to their gain ratio, and then removed, ten by ten, from the worst to the best (according to their rank), from the original dataset. Every time ten attributes were removed from the training set, the new dataset was mined by the three baseline algorithms following the sequential covering approach (i.e., both versions of CN2 and Ripper). Note that C4.5Rules was not considered during this process because, for most of the datasets, it showed an execution error and was not able to extract a set of decision rules from the built decision tree. This process of removing ten attributes from the training set was repeated until the predictive accuracy of at least one of the baseline algorithms dropped. In this case, the last ten removed attributes were added again to the training set and then removed, one by one, from the worst to the best, while the predictive accuracy of all the classifiers did not decrease.

The five datasets listed in Table 6.18 are related to two different application domains (corresponding to two types of proteins): protein involved in postsynaptic activity and G-Protein-Coupled-Receptor (GPCR) proteins. The creation of the postsynaptic dataset is described in detail in [18]. This dataset contains a class attribute indicating whether or not a protein has post-synaptic activity, and its predictor attributes consist mainly of "Prosite motifs" [10], where each motif is a type of "signature" present in some proteins having a certain function, as mentioned earlier. Identifying proteins with postsynaptic activities is of significant intrinsic interest because they are connected with the functioning of the nervous system.

In turn, identifying GPCR proteins and their families is particularly important for medical applications, since it is believed that 40–50% of current medical drugs target GPCR activity [8]. GPCRs are essentially transmembrane proteins (i.e., they cross the cell's membrane) that transmit signals received from outside the cell to within the cell. Different signals activate different biological processes within the cell, and GPCRs are involved in the transmission of many different types of signals. In the GPCR datasets, the class to be predicted is the function of the GPCR, which depends on which type of molecule binds to the part of the GPCR outside the cell and triggers the signal transmission into the cell. The predictor attributes vary depending on the dataset, but all the three types of motifs used as attributes (Prosite, Interpro, and Prints motifs [10]) represent protein signatures. The creation of the GPCR datasets is described in [11].

An important point to notice in the GPCR datasets is that the classes of GPCR proteins are organized into a hierarchical structure. Actually, these datasets were used to evaluate a hierarchical classification algorithm in [11]. Since hierarchical classification is out of the scope of this work, these datasets were "flattened." All the GPCR datasets, except GPCR-Prosite-L2, consider just the classes in the first, top level of the class hierarchy, ignoring lower-level classes. GPCR-Prosite-L2, in turn, considers the classes in the second level of the class hierarchy. Note that GPCR-Prosite-L2 is composed essentially of the same proteins and attributes as those found in GPCR-Prosite. However, GPCR-Prosite considers as class values only the nine classes present in the first hierarchical class level (ignoring lower-level

Table 6.19 Comparing the predictive accuracies (%) obtained by the GGP-RIs tailored to a specific bioinformatics dataset with selected attributes against the predictive accuracies (%) obtained by the baseline algorithms when using the complete dataset

Dataset	GGP-RIs	OrdCN2	UnordCN2	Ripper	C45Rules
Postsynaptic	98.32±0.24	98.7±0.22	98.42±0.12	98.3±0.22	97.82±0.32
GPCR-Prosite	86.44±0.13	87.34±0.32	75±0.2	87.66±0.39	82.64±2.44
GPCR-Prosite-L2	63.81±0.96	68.56±0.29	48.9±0.5	62.56±0.75	44±1.07
GPCR-Prints	90.71±0.69	91.04±0.25	83.84±1.52	91.06±0.27	88.94±2.76
GPCR-Interpro	89.34±0.29	90.88±0.45	90.56±0.41	90.26±0.32	80.36±6.35

classes), while GPCR-Prosite-L2 considers only the 50 classes in the second hierarchical class level (ignoring both the first-level classes and the classes below the second level). The differences in the number of examples when comparing these two datasets are due to proteins where only the first class in the hierarchy is known. Of course, these proteins are used as examples in GPCR-Prosite but not in GPCR-Prosite-L2.

Experiments with the GGP system were performed using the bioinformatics datasets with selected attributes only. Each dataset was divided into three subsets and inserted into the meta-training and meta-test sets of the GGP system, as described earlier. For these experiments, the comparisons of the GGP-RIs and the baseline algorithms were executed in two phases. First, the GGP-RIs' predictive accuracies were compared to the accuracies obtained by the baseline algorithms when run with the complete datasets (before any attribute selection process was applied). The purpose of this analysis was to show that the attribute selection process did not significantly reduce the predictive accuracies obtained by the baseline algorithms in comparison with using the full datasets. These results are reported in Table 6.19. In the second phase, comparisons among the GGP-RIs and the baseline algorithms run in the datasets with selected attributes were carried out. The results obtained are shown in Table 6.20.

Both Tables 6.19 and 6.20 show the names of the datasets used in the experiments, followed by the predictive accuracies obtained by the GGP-RIs in the datasets with selected attributes. The following columns show the predictive accuracies obtained by ordered CN2, unordered CN2, Ripper, and C4.5Rules for the complete datasets in Table 6.19, and for the preprocessed datasets (with selected attributes) in Table 6.20. Note that, for the bioinformatics datasets, the C4.5Rules algorithm was not able to extract rules from the original C4.5 tree in most of the experiments performed. In these cases, for the sake of completeness, the accuracy reported is the one obtained by the C4.5 decision tree model.

In Tables 6.19 and 6.20, cells in dark gray represent statistically significant wins of the GGP-RIs over the respective baseline method, whereas cells in light gray represent statistically significant wins of the baseline method over the GGP-RIs according to a paired two-tailed Student's t-test with significance level 0.05.

As shown in Table 6.19, the results obtained by the GGP-RIs with selected attributes are significantly better than the results obtained by baseline algorithms with

Table 6.20 Comparing the predictive accuracies (%) obtained by the GGP-RIs tailored to a specific dataset and the baseline algorithms when using the bioinformatics datasets with selected attributes

Dataset	GGP-RIs	OrdCN2	UnordCN2	Ripper	C45Rules
Postsynaptic	98.32±0.24	98.4 ± 0.2	98.4 ± 0.2	98.21 ± 0.22	98.4 ± 0.2
GPCR-Prosite	86.44±0.13	86.86±0.28	75.68±0.31	86.64±0.31	84.04±2.04
GPCR-Prosite-L2	63.81±0.96	68.38±0.65	47.98±0.1	62.36±0.42	NA
GPCR-Prints	90.71±0.69	90.78±0.32	90.64±0.37	91.05±0.14	86.76±2.58
GPCR-Interpro	89.34±0.29	90.2±0.25	88.46±0.55	89.83±0.32	84.02±3.73

the complete dataset in four cases, and significantly worse in three cases. The GGP-RIs are significantly better than unordered CN2 in three datasets, and significantly better than C4.5Rules in the dataset GPCR-Prosite-L2. At the same time, the GGP-RIs are significantly worse than ordered CN2 in GPCR-Prosite and GPCR-Prosite-L2, and significantly worse than Ripper in GPCR-Prosite.

If we compare these results with the ones reported in Table 6.20, we notice that now the GGP-RIs are significantly better than the baseline algorithms in two cases and significantly worse in only one. This is because the attribute selection process not only improved the accuracy of unordered CN2 in GPCR-Prints, but also decreased the accuracy of ordered CN2 and Ripper in GPCR-Prosite, making the corresponding previous wins and losses of the GGP-RIs not statistically significant anymore. Note that the predictive accuracy of C4.5Rules in the dataset GPCR-Prosite-L2 with selected attributes is not available. This is because C4.5 presented an error when it was being executed in this dataset and was not able to generate any classification model.

These results show that the GGP system is able to obtain statistically better results than unordered CN2 in two datasets, and competitive results in the other three datasets. At the same time, for the dataset GPCR-Prosite-L2, the results show that apparently the GGP system was not able to find ordered CN2, the algorithm with best performance among the four manually designed algorithms in Table 6.20, although ordered CN2 is included in the GGP system's search space. However, when analyzing the 25 GGP-RIs produced for GPCR-Prosite-L2, we observe that six of them were actually instances of ordered CN2. Results obtained by the GGP-RIs were always as good as the results obtained by Ripper and C4.5Rules for all the datasets. Further results on the postsynaptic dataset can be found in [19].

Again, based on these results, we can claim that the GGP system is able to produce competitive rule induction algorithms tailored to a specific real-world dataset or application domain.

6.3.4 A Note on the GGP System's Execution Time

In Section 6.2.8, we discussed the factors that influence the execution time of the GGP system, and explained why it is so difficult to estimate a best or worst runtime

Table 6.21 GGP system's runtime for experiments evolving algorithms tailored to a specific dataset

Meta-training set	Time (hr:min)	
	Best	Worst
Balance-scale	0:21	0:45
Postsynaptic	1:41	4:56
Lymphs	1:20	3:05
Promoters	15:37	43:18
GPCR-Interpro	63:26	169:30

for the GGP system. We then reported results for the best and worst case scenarios empirically obtained when running the proposed GGP system with different numbers of datasets in the meta-training set.

Similarly, Table 6.21 reports the same kind of information when running the GGP system with a single dataset in the meta-training set. The first column shows the name of the dataset in the meta-training set, followed by the best and worst runtimes, reported in the format hours:minutes. Out of the 25 datasets used in those experiments (i.e., 20 UCI datasets plus five bioinformatics datasets), where each dataset was used in a separate experiment, Table 6.21 reports the computation times for five datasets, chosen as follows: *balance-scale* and *postsynaptic* are the datasets associated with the fastest run of the GGP system among the UCI and bioinformatics datasets, respectively, while *promoters* and *GPCR-Interpro* are associated with the slowest GGP system run. The times reported for *lymphs* represent an average time for medium-sized UCI datasets. Again, all the experiments were performed on Pentium 4 duo processor machines with 1 GB RAM and running Linux.

6.4 Summary

This chapter presented the results obtained by the GGP system proposed in Chapter 5 in two scenarios: (a) when evolving rule induction algorithms robust across different application domains; and (b) when evolving rule induction algorithms tailored to one specific application domain or dataset. In scenario (a), experiments were performed with 20 UCI datasets. In scenario (b), experiments were first performed with the same 20 UCI datasets, and later with a set of five real-world bioinformatics datasets, involving the prediction of protein functions.

For all the experiments reported in this chapter, the predictive accuracies obtained by the GGP-RIs were compared to the predictive accuracies of four well-known, human-designed rule induction algorithms, namely ordered CN2, unordered CN2, Ripper, and C4.5Rules. In general, the results showed that the GGP-RIs were competitive with these baseline algorithms. In addition, we have shown that the proposed GGP system produced, overall, better rule induction algorithms (i.e., algorithms with higher predictive accuracy) than a grammar-based hill-climbing system. We have also shown the pseudocode of some of the evolved GGP-RIs, and highlighted their innovative features.

References

1. Banzhaf, W., Nordin, P., Keller, R.E., Francone, F.D.: Genetic Programming – An Introduction; On the Automatic Evolution of Computer Programs and its Applications. Morgan Kaufmann (1998)
2. Brodley, C., Friedl, M.: Identifying mislabeled training data. Journal of Artificial Intelligence Research **11**, 131–167 (1999)
3. Clare, A., King, R.D.: Machine learning of functional class from phenotype data. Bioinformatics **18**(1), 160–166 (2002)
4. Clark, P., Boswell, R.: Rule induction with CN2: some recent improvements. In: Y. Kodratoff (ed.) Proc. of the European Working Session on Learning on Machine Learning (EWSL-91), pp. 151–163. Springer-Verlag, New York, NY, USA (1991)
5. Clark, P., Niblett, T.: The CN2 induction algorithm. Machine Learning **3**, 261–283 (1989)
6. Cohen, W.W.: Fast effective rule induction. In: A. Prieditis, S. Russell (eds.) Proc. of the 12th Int. Conf. on Machine Learning (ICML-95), pp. 115–123. Morgan Kaufmann, Tahoe City, CA (1995)
7. Correa, E.S., Freitas, A.A., Johnson, C.G.: A new discrete particle swarm algorithm applied to attribute selection in a bioinformatics data set. In: Proc. of the Genetic and Evolutionary Computation Conf. (GECCO-06), pp. 35–42. ACM Press (2006)
8. Fillmore, D.: It's a GPCR world. Modern Drug Discovery **11**(7), 24–28 (2004)
9. Freitas, A.A., Wieser, D., Apweiler, R.: On the importance of comprehensible classification models for protein function prediction. IEEE/ACM Transactions on Computational Biology and Bioinformatics (in press)
10. Higgs, P.G., Attwood, T.K.: Bioinformatics and Molecular Evolution. Blackwell (2005)
11. Holden, N., Freitas, A.: Hierarchical classification of G-protein-coupled receptors with a PSO/ACO algorithm. In: Proc. of the IEEE Swarm Intelligence Symposium (SIS-06), pp. 77–84. IEEE Press (2006)
12. J.He, Yao, X.: Towards an analytic framework for analyzing the computation time of evolutionary algorithms. Artificial Intelligence **145**(1-2), 59–97 (2003)
13. Karwath, A., King, R.: Homology induction: the use of machine learning to improve sequence similarity searches. BMC Bioinformatics **3**(11), online publication (2002). http://www.pubmedcentral.nih.gov/articlerender.fcgi?artid=107726
14. Liu, H., Motoda, H. (eds.): Feature Selection for Knowledge Discovery and Data Mining. Kluwer (1998)
15. Mirkin, B., Ritter, O.: A feature-based approach to discrimination and prediction of protein folding groups. In: Genomics and Proteomics, pp. 155–177. Springer (2000)
16. Newman, D.J., Hettich, S., Blake, C.L., Merz, C.J.: UCI Repository of machine learning databases. University of California, Irvine, http://www.ics.uci.edu/~mlearn/MLRepository.html (1998)
17. Pappa, G.L.: Automatically evolving rule induction algorithms with grammar-based genetic programming. Ph.D. thesis, Computing Laboratory, University of Kent, Canterbury, UK (2007)
18. Pappa, G.L., Baines, A.J., Freitas, A.A.: Predicting post-synaptic activity in proteins with data mining. Bioinformatics **21**(Suppl. 2), ii19–ii25 (2005)
19. Pappa, G.L., Freitas, A.A.: Automatically evolving rule induction algorithms tailored to the prediction of postsynaptic activity in proteins. Intelligent Data Analysis **13**(2), 243–259 (2009)
20. Pappa, G.L., Freitas, A.A.: Evolving rule induction algorithms with multi-objective grammar-based genetic programming. Knowledge and Information Systems **19**(3), 283–309 (2009)

21. Pazzani, M.J.: Knowledge discovery from data? IEEE Intelligent Systems **15**(2), 10–13 (2000)
22. Quinlan, J.R.: C4.5: programs for machine learning. Morgan Kaufmann (1993)
23. Russell, S., Norvig, P.: Artificial Intelligence: A Modern Approach. Prentice Hall (2002)
24. Szafron, D., Lu, P., Greiner, R., Wishart, D., Poulin, B., Eisner, R., Lu, Z., Poulin, B., Anvik, J., Macdonnel, C.: Proteome analyst – transparent high-throughput protein annotation: function, localization and custom predictors. Nuclei Acids Research **32**, W365–W371 (2004)
25. Witten, I.H., Frank, E.: Data Mining: Practical Machine Learning Tools and Techniques with Java Implementations, 2nd edn. Morgan Kaufmann (2005)
26. Wong, M.L., Leung, K.S.: Data Mining Using Grammar-Based Genetic Programming and Applications. Kluwer, Norwell, MA, USA (2000)

Chapter 7
Directions for Future Research on the Automatic Design of Data Mining Algorithms

In the previous chapters we have presented a new Grammar-based Genetic Programming (GGP) system to automatically design rule induction algorithms. The GGP system works with a grammar that contains background knowledge about how human experts design rule induction algorithms and some other interesting components that, to the best of our knowledge, have not been used by human-designed rule induction algorithms so far. The results of extensive computational experiments showed that the GGP system can effectively evolve rule induction algorithms using two different approaches:

- The evolution of *robust* rule induction algorithms, which perform well in a variety of application domains different from the application domains (datasets) used during the GGP system's run.
- The evolution of rule induction algorithms *tailored to a single specific application domain*, where the data subset used during the GGP system's run belongs to the same application domain to which the evolved rule induction algorithm will be applied.

Note that, in both approaches, the evolved rule induction algorithms are applied to test data unseen during training, as usual in the classification task of data mining.

The predictive accuracies obtained by the rule induction algorithms evolved by the GGP system (GGP-RIs) on the test data were shown to be competitive with the predictive accuracies obtained by well-known, manually designed (and refined over decades of research) rule induction algorithms. The experimental results also showed that overall the GGP-RIs obtained significantly higher predictive accuracies than the rule induction algorithms produced by a Grammar-based Hill Climbing (GHC) method. These results were produced in controlled experiments where both the GGP system and the GHC system used the same grammar, the same fitness (evaluation) function, and the same individual representation, and evaluated the same number of candidate rule induction algorithms during their search. Hence, one can conclude that the GGP system was considerably more effective than the GHC system.

G.L. Pappa, A.A. Freitas, *Automating the Design of Data Mining Algorithms*,
Natural Computing Series, DOI 10.1007/978-3-642-02541-9_7,

An analysis of the evolved GGP-RIs showed that, besides being competitive with human-designed algorithms, many of them present some innovative way of refining rules and/or integrating pre- and post-pruning techniques. Experiments also revealed that, when evolving *robust* rule induction algorithms, the GGP system takes a more conservative approach and builds simpler algorithms. In contrast, when generating GGP-RIs *tailored to a specific application domain*, the GGP system evolves more innovative algorithms, in the sense of algorithms with some procedures different from those of human-designed rule induction algorithms.

Nevertheless, the GGP system presented here represents only the first step towards the automation of the design of data mining algorithms, and leaves plenty of room for several improvements and future research directions. This chapter discusses some of these future directions.

First, Section 7.1 presents some potential improvements to the GGP system proposed in Chapter 5. Next, Section 7.2 introduces a third approach the GGP system could use, namely, evolving rule induction algorithms tailor-made for groups of datasets with similar characteristics. Section 7.3 discusses which other types of search methods could be used for automatically designing data mining algorithms. Finally, Sections 7.4 and 7.5 identify other types of classification algorithms and data mining algorithms, respectively, that could be automatically designed.

7.1 Potential Improvements to the Current GGP System

This section suggests two modifications that could potentially improve the performance of the proposed GGP system. These modifications regard the system's two most important components: the grammar and the fitness.

7.1.1 Improving the Grammar

The grammar is the element of the GGP system that determines the search space, i.e., the set of rule induction algorithms that can be designed by the GGP system. We believe that there are some extensions that, if inserted into the current grammar, could lead the GGP system to evolve even more original and innovative rule induction algorithms. There are in particular five extensions that could be introduced:

- Include in the nonterminal *EvaluateRule* grammar production rules similar to the ones proposed in Wong [22], which generate the evaluation function of rule induction algorithms. The new system would then be able to automatically construct (in a data-driven way) new rule evaluation heuristics, rather than being limited to select one out of predefined rule evaluation heuristics such as confidence, Laplace estimation, information content, and information gain. This could lead to new rule induction algorithms with rule evaluation heuristics tailored to the dataset at hand.

- Include other types of rule representation in the grammar, such as first-order logic. The current grammar includes only propositional rule representation. First-order logic has greater expressiveness power than propositional logic, but on the other hand the size of the search space of the GGP system would be considerably greater. Hence, the pros and cons of these two kinds of rule representation need to be investigated in the context of the proposed GGP system.
- Insert into the grammar more complex and/or innovative components of rule induction algorithms, such as the minimum description length [17] heuristic, used by Ripper, or measures of rule interestingness, as described in Section 2.4.3.
- Extend the grammar to be adaptable, based on the performance of the best individuals at each generation, in a way similar to that in the work of Whigham [21].
- Use an approach such as the GE^2 (Grammatical Evolution Squared) [16], where the grammar used to represent the background knowledge and syntactic restrictions of the target problem is itself evolved by a grammatical evolution algorithm. In this case, an evolution of the genetic code per se is performed, as the search space evolves together with the solutions.

7.1.2 Modifying the GGP System's Fitness Function

After the grammar, the element which has more impact in the search mechanism of the GGP system is its fitness function, which evaluates the quality of candidate rule induction algorithms. During the development of the system, we compared the results obtained when using three types of fitness functions, accuracy, sensitivity × specificity (see Eq. 5.3), and the current fitness of the system, based on a variation of accuracy, adjusted to take into account the frequency of the majority class (see Eq. 5.5).

Nevertheless, in the last few years, another approach for evaluating the effectiveness of a classification algorithm has been widely disseminated: the area under the ROC curve (AUC) [5, 6]. A comparison of the results of the GGP system with the current fitness function and the AUC would be interesting.

Furthermore, the current fitness function could also be extended to take into account the efficiency (processing time) of the candidate rule induction algorithms, as in the GP systems proposed by [7] and [1] for automating the design of heuristics for a combinatorial optimization algorithm.

In addition, the issue of multiobjective optimization should be further investigated. In Section 6.2.7 we reported results of a multiobjective version of the fitness function, based on Pareto dominance, but in the future it would be interesting to use a lexicographic approach instead of a Pareto dominance-based approach to implement a multiobjective fitness function. As discussed in Section 3.3.2, the lexicographic approach, which essentially consists of ranking objectives in decreasing order of priority to the user and optimizing them in that order, seems naturally suitable for problems where the user clearly considers one objective more important than another, even though the user would find it difficult and unnatural to assign numerical weights (indicators of relative importance) to different objectives. For instance, in the classification task of data mining, most researchers would consider that maximizing the predictive accuracy of the algorithm is more important than minimizing the size of the classification model, important information that would be ignored by the Pareto dominance-based approach. The lexicographic approach has been much less investigated than the Pareto dominance-based approach, in the areas of both data mining and evolutionary algorithms, so there is a clear need for more research on the lexicographic approach in these areas.

7.2 Designing Rule Induction Algorithms Tailored to a Type of Dataset

Recall that Chapter 6 presented two approaches in which the GGP system was able to evolve rule induction algorithms. In the first one, we generated *robust* rule induction algorithms, which were designed to be effectively applied to any classification dataset, regardless of the application domain. In the second one, rule induction algorithms *tailored to a specific application domain* were generated.

A third approach that could be considered is to use the proposed GGP system to produce rule induction algorithms *tailored to a type of dataset*, i.e., a group of datasets with similar characteristics, where that group would possibly contain datasets from different application domains. In this approach, datasets would be grouped according to some common properties, and only datasets belonging to a given group would be allowed in the GGP system's meta-training set. Note that this third approach represents an intermediate case between the two extreme cases of training the GGP system with a number of datasets from unrelated application domains (to evolve robust rule induction algorithms) and with data from a single application domain (to evolve algorithms tailored to that domain).

For instance, in the bioinformatics field, there are a lot of datasets which, although derived from different application domains, share several important characteristics from a data mining viewpoint. In particular, there are many bioinformatics datasets that contain a very large number of binary attributes, very sparse data, and very unbalanced classes. The postsynaptic dataset used in Section 6.3.3 is an example of such datasets. Other examples are found in [9, 10].

It is possible to target a group of datasets with similar characteristics when producing rule induction algorithms, although in practice this is a difficult problem. The main difficulty is how to measure the degree of similarity between different datasets from different application domains. Dataset characterization [3, 13, 15] is in general an open problem in the meta-learning literature (despite significant progress of the research in this area), but we think it is an interesting research direction.

7.3 Investigating Other Types of Search Methods for Automated Algorithm Design

The GGP system proposed in Chapter 5 is the first, to the best of our knowledge, to design a complete rule induction algorithm. Hence, to evaluate its performance, we chose to compare it with a simple greedy hill-climbing search method. However, we make no claim the proposed GGP system is the best type of method for automatically designing data mining algorithms.

It would be interesting to investigate the use of other types of methods capable of performing the same task. Here we point out three other types of evolutionary methods that could be implemented. The first of them consists of using the grammatical evolution approach discussed in Section 3.6.3. The second consists of trying out evolutionary approaches based on probability theory, such as Estimation Distribution Algorithms (EDAs) [11, 14]. The third method is a combination of the former two, where a kind of hybrid EDA-GP based on a grammar is used as a search method [18].

Concerning the aforementioned first type of method, when choosing which type of grammar-based GP to use in this research, we did not find any significant evidence that showed that either GGP using solution-encoding individuals or GGP using production-rule-sequence-encoding individuals was superior to the other. Hence, we decided to use the solution-encoding individual representation because we considered it simpler (see Section 5.1 for a detailed explanation).

However, it would be interesting to combine the grammar developed in this work with a grammatical evolution system following the production-rule-sequence-encoding individual representation, and to compare both systems' behaviors and results.

A rather different type of method to examine would be EDAs. An EDA is a type of evolutionary algorithm where crossover and mutation operations are nonexistent. Instead, the algorithm works by building probabilistic models of the relationships between the components of candidate solutions. The main role of the population of individuals is to provide information to create such probabilistic models. In EDAs, the individuals in a given generation are evaluated according to a fitness function and undergo a selection process similar to what happens in conventional evolutionary algorithms. However, after that, the selected individuals are used to update the probabilistic model being built, rather than to produce offspring. After this update, the current population is discarded, and a new population is sampled from the probabilistic model. This process is iteratively repeated for a predefined number of iterations or until a satisfactory solution is found.

When using EDAs, one does not need to worry about which crossover or mutation methods to use, about how often crossover and mutation should be applied, and so on. In addition, EDAs have the advantage of using a mathematically-sound method, based on probability theory, to support the evolutionary process. It should be noted, however, that the price paid for avoiding genetic operators and related parameters is that one needs to specify instead parameters related to the probabilistic model being built along the evolution, e.g., whether the components of the probabilistic model should have their probabilities computed independently of each other (in a univariate fashion) or in a multivariate fashion. In any case, EDAs and related probabilistic model building algorithms seem interesting and relatively less explored types of evolutionary algorithms for program induction [8], and it would be interesting to evaluate their effectiveness in the automated design of data mining algorithms.

Another alternative is to use a kind of hybrid EDA-GP system as the one proposed in [18], where a stochastic context-free grammar (CFG) is used to represent the probabilistic model being evolved. In a stochastic CFG, each production rule is associated with a probability, which is taken into account when creating a new population of individuals. During the evolution of the GP algorithm, both the structure and/or the probabilities associated with the production rules are learned, and production rules can be split or merged to reflect the fittest individuals. A more detailed review on hybrid EDA-GPs can be found in [19].

In addition, as mentioned in the Introduction (Section 1.2.1), it would be interesting to develop an Inductive Logic Programming (ILP) system for automatically designing a full data mining algorithm, since ILP is a large research area at the intersection of data mining and automated program synthesis [2, 4, 12].

7.4 Automatically Designing Other Types of Classification Algorithms

The work presented in this book showed it is possible to automatically design an effective rule induction algorithm. We chose to focus on the design of this type of classification algorithm for two main reasons. First, because it generates human-comprehensible classification models, which can be interpreted by users in order to try to get new insight into the data or the application domain (for a review of the motivation for discovering comprehensible knowledge, see Section 2.2.3). Secondly, because we observed a natural "evolution" of rule induction algorithms over the last 30 years of research in the area. Indeed, most rule induction algorithms are usually obtained from the combination of a basic algorithm (typically following the sequential covering approach) with new evaluation functions, pruning methods, and stopping criteria for refining or producing rules gradually, generating many "new" and more sophisticated sequential covering algorithms. Hence, it was natural to take advantage of this "human-driven evolution" and extend it to a new type of evolution, more precisely, an artificial (and much faster) evolution, using a grammar-based genetic programming system.

However, of course these motivations are not restricted to rule induction algorithms, and many other types of classification algorithms could also be automatically designed. Here we give examples of two other types of classification algorithms capable of generating human-comprehensible knowledge: decision tree induction and Bayesian networks. The design of decision tree induction algorithms, as described in Section 2.3, shares similarities with the design of rule induction algorithms. This is because decision tree induction algorithms also follow a basic algorithm (which is usually based on the top-down approach) whose design also involves choices among different evaluation and stopping criteria and pre-pruning and post-pruning strategies.

Bayesian networks, in contrast, are graphical models that represent probabilistic relationships among a set of variables. They also have the advantage of representing comprehensible knowledge. When using a grammar-based approach as the one described in Chapter 5, a decision tree induction algorithm or a Bayesian network algorithm could be automatically designed by changing the grammar to include in its production rules components or building blocks of the respective type of algorithm. Note that, although the current grammar would have to be completely replaced by a grammar about another type of classification algorithm, the part of the GGP system referring to the GP search would not need any significant modification in principle.

7.5 Automatically Designing Other Types of Data Mining Algorithms

The focus of this book is on the design of algorithms to solve the classification task of data mining. However, using the same principles, we could automatically design algorithms to tackle other data mining tasks, such as clustering and regression.

Let us consider the clustering task. Clustering is commonly an unsupervised learning task, where we want to create clusters of unlabeled examples according to their relative proximity in the data space. That is, a cluster should consist of examples that are somehow close to each other in the data space. There are many types of clustering algorithms, including hierarchical, partitional, and density-based clustering [20]. While hierarchical clustering iteratively merges smaller clusters into larger ones or splits larger clusters into smaller ones, producing a hierarchy of clusters, partitional clustering divides the dataset into a flat set of (overlapping or non-overlapping) clusters. Density-based clustering algorithms, as the name suggests, create clusters according to variations in the density of examples in different areas of the data space.

Given the large variety of algorithms that can be created, in order to automate the design of clustering algorithms it seems we would need to focus on one type of clustering algorithm at first (as we did with the classification task), to make the problem more tractable. Suppose we choose to automatically design partitional clustering algorithms. First, we need to identify the main components of these algorithms, such as the intra- and inter-clustering dissimilarity functions, as we did in the rule learning approach. Assuming we use a grammar to incorporate background knowledge about the design of clustering algorithms, the main components of the chosen type of clustering algorithm should be represented by a grammar, which could be then used by a GGP system (or any other new type of method proposed for the same purposes) to automatically design clustering algorithms.

References

1. Bader-El-Den, M., Poli, R.: Generating SAT local-search heuristics using a GP hyper-heuristic framework. In: Artificial Evolution (Proc. of 8th Int. Conf. on Evolution Artificielle), *LNCS*, vol. 4926, pp. 37–49. Springer-Verlag (2007)
2. Bergadano, F., Gunetti, D.: Inductive Logic Programming: from machine learning to software engineering. MIT Press (1996)
3. Brazdil, P., Giraud-Carrier, C., Soares, C., Vilalta, R.: Metalearning: applications to data mining. Springer (2009)
4. Dzeroski, S., Lavrac, N. (eds.): Relational Data Mining. Springer (2001)
5. Fawcett, T.: ROC graphs: notes and practical considerations for data mining researchers. Tech. Rep. HPL-2003-4, HP Labs (2003)
6. Flach, P.: The geometry of ROC space: understanding machine learning metrics through ROC isometrics. In: Proc. 20th Int. Conf. on Machine Learning (ICML-03), pp. 194–201. AAAI Press (2003)
7. Fukunaga, A.: Automated discovery of composite SAT variable-selection heuristics. In: Proc. of the National Conf. on Artificial Intelligence (AAAI-02), pp. 641–648. AAAI Press (2002)
8. Hasegawa, Y., Iba, H.: A Bayesian network approach to program generation. IEEE Transactions on Evolutionary Computation **12**(6), 750–764 (2008)
9. Holden, N., Freitas, A.: Hierarchical classification of G-protein-coupled receptors with a PSO/ACO algorithm. In: Proc. of the IEEE Swarm Intelligence Symposium (SIS-06), pp. 77–84. IEEE Press (2006)
10. Holden, N., Freitas, A.A.: A hybrid particle swarm/ant colony algorithm for the classification of hierarchical biological data. In: P. Arabshahi, A. Martinoli (eds.) Proc. of the IEEE Swarm Intelligence Symposium (SIS-05), pp. 100–107. IEEE (2005)
11. Larranaga, P., Losano, J. (eds.): Estimation of Distribution Algorithms. Kluwer (2002)
12. Lavrac, N., Dzeroski, S.: Inductive Logic Programming: techniques and applications. Ellis Horwood (1994)
13. Lindner, G., Studer, R.: Ast: Support for algorithm selection with a CBR approach. In: Proc. of the 3rd European Conf. on Principles and Practice of Knowledge Discovery in Databases (PKDD-99), pp. 418–423. Springer-Verlag, London, UK (1999)
14. Lozano, J., Larranaga, P., Inza, I., Bengoetxea, E. (eds.): Towards a New Evolutionary Computation: advances in estimation of distribution algorithms. Springer (2006)
15. Michie, D., Spiegelhalter, D.J., Taylor, C.C., Campbell, J. (eds.): Machine learning, neural and statistical classification. Ellis Horwood, Upper Saddle River, NJ, USA (1994)
16. O'Neill, M., C.Ryan: Grammatical evolution by grammatical evolution: The evolution of grammar and genetic code. In: In Proc. of European Conf. on Genetic Programming (EuroGP-04), pp. 138–149 (2004)
17. Quinlan, J.R.: Learning logical definitions from relations. Machine Learning **5**, 239–266 (1990)
18. Shan, Y., McKay, R.I., Abbass, H.A., Essam, D.: Program distribution estimation with grammar models. Complexity International **11**, 191–205 (2005)
19. Shan, Y., McKay, R.I., Essam, D., Abbass, H.A.: A survey of probabilistic model building genetic programming. Tech. Rep. TR-ALAR-200510014, University of New South Wales (2005)
20. Tan, P., Steinbach, M., Kumar, V.: An Introduction to Data Mining. Addison-Wesley (2006)
21. Whigham, P.A.: Grammatical bias for evolutionary learning. Ph.D. thesis, School of Computer Science, University College, University of New South Wales, Australian Defence Force Academy, Canberra, Australia (1996)
22. Wong, M.L.: An adaptive knowledge-acquisition system using generic genetic programming. Expert Systems with Applications **15**(1), 47–58 (1998)

Index

G.L. Pappa, A.A. Freitas, *Automating the Design of Data Mining Algorithms*,
Natural Computing Series, DOI 10.1007/978-3-642-02541-9,